GROWING
WONDER

Publisher: BLOOM Imprint

Author: FELICIA ALVAREZ

Contributing Photographers: JILL CARMEL, ASHLEY LIMA, ASHLEY NOELLE EDWARDS

Editorial Director: DEBRA PRINZING

Creative Director: ROBIN AVNI

Designer + Ilustrator: JENNY MOORE-DIAZ

Copy Editor: BRENDA SILVA

Cover Photography: JILL CARMEL

Copyright (c) 2022 by BLOOM Imprint

All rights reserved. No part of this publication may be reproduced, stored in a retrieval system or transmitted, in any form or by any means, electronic, mechanical, photocopying, recording or otherwise, without prior written permission of the publisher.

ISBN: 978-1-7368481-2-8

Library of Congress Control Number: 2022931431

BLOOM Imprint
4810 Pt. Fosdick Drive NW, #297, Gig Harbor, WA 98335
www.bloomimprint.com

Printed in the U.S.A. by Consolidated Press, Seattle, WA

A FLOWER FARMER'S GUIDE TO ROSES

GROWING WONDER

FELICIA ALVAREZ
MENAGERIE FARM & FLOWER

For my family

You've always shown me how to grow with love

TABLE OF CONTENTS

12
INTRODUCTION

1
ARE ROSES DIFFICULT?
18

22
ROSE REALITIES

26
THE ROSE PRIMER

27
LIFECYCLE OF A ROSE

GLOSSARY

2
YOUR ROSE PERSONALITY
28

31
THREE ROSE PERSONALITIES

WEEKEND WARRIOR

EVERYDAY GARDENER

ASPIRING ROSARIAN

3
THE ROSE GALLERY
34

38
WHITE & CREAM

41
BLUSH

42
LIGHT PINK

44
DEEP PINK

47
BURGUNDY & WINE

48
LAVENDER & PURPLE

50
CRIMSON & RED

52
GOLDEN & BUTTER

54
PEACH & COPPER

56
MULTI & TAUPE

4
WHERE, WHEN & HOW TO PLANT ROSES
58

60
SELECTING A TYPE OF ROSE

66
WHERE & WHEN TO PLANT

72
TIME TO PLANT

78
PLANTING A BARE ROOT ROSE

80
ALSO NICE TO KNOW

81
PLANTING A POTTED ROSE

84
TRANSPLANTING A ROSE

5
YEAR-ROUND ROSE CARE
86

89
CARE PROGRAM OVERVIEW

90
DETERMINING THE TYPE OF CARE

92
MULCHING & COMPOSTING

95
FERTILIZING

98
PEST & DISEASE CONTROL

108
WEEDS

109
WATERING & IRRIGATION

114
PRUNING

122
MONTHLY CARE CHART

6
HARVEST & POST-HARVEST CARE
124

128
GARDEN ROSES VS. GREENHOUSE ROSES

129
PROPERTIES OF A GOOD CUT-GARDEN ROSE

131
ROSE WISDOM

132
SANITATION

133
THE ROSE HARVEST

138
PRESERVING YOUR ROSES

140
DESIGNING WITH ROSES

146
IN CLOSING

148
RESOURCES

PHOTOGRAPHY BY ASHLEY LIMA

INTRODUCTION

ROSES
ARE MY LOVE LANGUAGE

With a sweet fragrance wafting through the air and loamy soil beneath my feet, I trundled behind my mother, distracted as my eyes followed a gently buzzing bee emerge from soft petals after a snack. I was a master dawdler, always sticking my nose deep into another bloom to take in the intoxicating scent of a rose. "Hurry up, Felicia. I need my best little helper," my mother would call, as I scrambled to follow her with a bucket as she clipped roses from the garden on our farm.

In my Strawberry Shortcake dress, Holly Hobby bows, and curls bouncing up and down, I ran as fast as I could to catch up with her, while simultaneously picking each petal, one-by-one, off a bloom to throw to the wind. I was always greeted with a sweet kiss, pat on the forehead, and tug on my pigtails from her—especially when I arrived dirty and disheveled with smile on my face.

I would spend hours with my mother and my grandmother tending to the garden on our farm. During these times, I was the ultimate Chatty Cathy asking them a million questions such as: Where does the pollen go? Why are aphids so sticky? Why is this leaf brown? Why do the roses all have different scents? Who decides what color they will be? My mother would always answer her inquisitive future scientist, patiently showing me step-by-step what to do, and explaining how and why.

I was absolutely enamored with the wonder of roses. Of all the plants and trees in our farm's ecosystem, they were my favorite. I remember feeling an immense sense of pride and satisfaction after a morning harvest as I slid the step stool to the kitchen sink, tiptoed to fill a vase, and placed my freshly picked bounty of roses on our kitchen table. I had a hand in making this beautiful display of magic happen.

PREVIOUS PAGE. Nestled at the base of the Sutter Buttes, in the heart of the Sacramento Valley, Menagerie Farm & Flower is home to fruit trees, specialty cut flowers and thousands of my beloved garden roses.

PHOTOGRAPHY BY JILL CARMEL

INTRODUCTION

The farm garden, and roses in particular, were a passion project started by my grandmother. Along with my grandfather, she filled our family garden with a true menagerie of annual, perennial, and edible trees and plants—camellias, tulips, gladiolas, tomatoes, zucchini, petunias, pansies, freesias, lilies, mandarins, peaches, walnuts, lavender—and so much more. She passed her love of gardening along to my mother, and then to me as well.

We were three generations of women who loved to squish their toes into the soil among the birds, bees, flowers, and the trees. Watching these two women on my daily garden adventures shaped my relationship with the land, which became one of nurture, care, respect, and admiration. I cherish these vivid memories of a special time and place spent with them, not knowing they were setting the foundation for what my passion in life would become.

I grew up in the warm Sacramento Valley of California, at the base of the Sutter Buttes on a prune farm stewarded by my maternal grandparents. They grew many different crops over the years, from peaches and beans to canning tomatoes, rice, and almonds. I started working on the farm at an early age. My first "official" paid job (at eight-years-old) was hand-harvesting prunes from immature trees that were too young to be mechanically harvested. I was paid by the "lug,"—a large rectangular wooden box filled with the fruit. My sticky, prune juice-covered hands would pick as fast as I could as I tried to beat the adults on the crew. With my pigtails and Barbie T-shirts, they didn't see spirited little me coming. I always hoped to win the unofficial daily contest for the most lugs picked, and I even beat the adults a few times.

My grandfather mentored me in all aspects of farming, as I spent the rest of my childhood and early adulthood working in different roles: harvesting, processing, bookkeeping, and running our commercial prune dehydrator. It eventually came time for me to choose a path in life, and I knew I wanted a career in agriculture. I studied crop science and business in college, with a special fascination for entomology, plant pathology, and soil science. I was the uber-hip college gal with an insect collection on her dining table, and a roommate who asked what the frozen bugs were doing in the freezer when she pulled out a frozen pizza.

After graduation from Cal Poly San Luis Obispo, I spent my early career working in wineries and vineyards, and later opened a winery with my husband, Jeff Munsey. I eventually reached a plateau and knew it was time for something more meaningful for my life. What did that mean for me? It meant buying the family

INTRODUCTION

farm and returning to my first home. Growing prunes was rewarding, but I was itching to do something different. I still felt that voice inside say, "There's more for you out there."

Every spare moment of free time I had, which was not a lot with a newborn child and a farm to run, I was back on the land with my son in his stroller next to me as I deadheaded spent blooms in my family's old rose garden. They were calling me back: The bright-eyed, pigtailed girl who was where she felt most at home. This is how and when Menagerie Farm & Flower was born.

I started selling my cut garden roses to a few local floral designers, and as they say, the rest is history. Now, my farm has 72 acres of prunes, 20 acres of rice, three acres of flourishing garden roses, 1/4-acre of assorted specialty cut flowers, and a

> It's such a joy to be able to send the beauty of flowers and plants out into the world, and know the recipients are celebrating all of life's special moments.

nursery specializing in rose plants. In the winter, the coolers are stocked with bare root rose plants ready to make their way to their forever homes on gardens and farms across the county. Come early spring, the farm is filled with fresh-cut tulips and flowering branches.

My collection of garden roses includes modern hybrid teas, floribundas, English varieties, and climbers—each with special characteristics—as they come alive in late spring. Professional floral designers covet my cut garden roses for weddings and special events, while flower-lovers from across the country purchase fresh-picked bouquets of garden roses that I ship right from the farm to their door.

The spring is also the beginning of potted rose season. Rose-lovers from near and far make the pilgrimage to our on-farm nursery to pick up roses for their own farms and gardens. It's such a joy to be able to send the beauty of flowers and plants out into the world, and know the recipients are celebrating all of life's special moments with the harvest of Menagerie Farm's flowers.

However, the most rewarding and enjoyable part of my farm evolution by far is mentoring the young, and young-at-heart, farmers and gardeners. These are the

INTRODUCTION

rose-lovers who attend workshops on the farm; join my online learning community, the Menagerie Academy; and follow our flowers and inspiration through the wonderful world of social media. My approach to teaching is simple: I blend my foundation in agricultural science with 20-plus years of farming experience to bridge the gap between academia and the real-world conditions of a garden or farm. I believe my in-the-trenches knowledge is both approachable and practical for today's garden rose enthusiast and modern flower farmer.

Little did I know that my path through life began in the garden at an early age with my mother, and later with mentoring from my grandfather. As my path continued with studying agriculture in college, and eventually running my own winery, my chosen path came full circle and returned me to the helm of the family orchards. Looking back, every step provided me with the necessary building blocks to ultimately create Menagerie Farm & Flower.

> Wherever you are in your gardening or farming journey, it's right where you need to be right now.

However, my journey, like many others, was not a linear one. It had its share of zigs and zags, triangles and squares, along the way. So, wherever you are in your gardening or farming journey, it's right where you need to be right now. Whether you have a small plot in the backyard, or a single pot on a patio or acres of land, I am going to teach you. My step-by-step instruction will explain how to grow garden roses, while respecting the land and all the life it has to give—taught to you in the same way my mother taught it to me many years ago. All you need to bring is a heart open to learning, a simple vase, and a table to enjoy your bounty.

Today, as I take walks at sunset with my husband and two young sons in tow, my family ancestors are still here with me in spirit as we all meander through the paths of our old garden that wind into my new garden and acres of roses. As I smell the calming perfume that wafts across the fields, I truly appreciate this beautiful land that I am privileged to steward. I know I am finally doing what I am meant to do: growing, sharing, and inspiring others to find their own path to growing wonder—not just garden roses but wonderful dreams too.

So welcome, my friends, to the Menagerie Farm family, I'm happy to have you here. Let's walk together now and make rose magic happen.

Rose harvesting is a daily farm practice that gives me immense joy. Here, my bucket is filled with fresh-picked 'Honey Dijon' roses.

PHOTOGRAPHY BY JILL CARMEL

1

IN THIS CHAPTER

ROSE REALITIES

THE ROSE PRIMER

GLOSSARY

ARE ROSES DIFFICULT?

With more than 20 years of professional farming under my belt—I've grown everything from French Prunes and wine grapes, to rice and canning tomatoes—I'm continually asked more questions about growing roses than any other plant. As a flower farmer who specializes in field-grown garden roses, the number-one question I'm asked is, "Are roses difficult to grow?" It's likely a question you've asked at some point in your rose-growing journey, as well.

Well, I'm here to answer that question and dispel the assumption that roses are finicky, hard-to-grow plants. I hope that you, like me, will discover a true affection for roses and a confidence to grow them.

Roses are my love language, really. My grandmother and mother were amateur rose collectors. I spent my childhood following them around, deadheading stems and smelling the blooms. I truly fell in love with roses at an early age. Garden roses were a part of the fabric of our simple farm garden, and I learned to appreciate their distinct qualities. I eventually even appreciated that all of the thorns, pests, and diseases that come along with roses are part of the greater farm ecosystem that ultimately creates a whole symphony of beauty. Roses are perfect in their imperfection and an amazing part of the sensory experience of my childhood, which I cherish today.

I've learned—from a rose-filled childhood to my role today as a professional farmer—that there is no one-size-fits-all approach to growing roses. Rather, it's all about selecting the best rose varieties for the right location. You should consider your garden's climate, cultural conditions, growing goals, and the time you have in your daily schedule to devote to care. When

A bucket holds recently harvested 'Distant Drums' roses.
The small white vase contains 'Abbaye de Cluny' peach roses, with a selection of rich red roses in the glass vessel, including 'Black Baccara' and 'Rouge Royale'.

PHOTOGRAPHY BY ASHLEY LIMA

searching for a good cutting rose to grow at production scale, it's easy for me to fall down the internet rabbit hole searching for that special, rare "unicorn" old rose—the completely unique one that has a perfect color or unforgettable fragrance. Or, I might have heard rumors of the next "it" rose variety that has been resurrected from obscurity. Inevitably, once I find what I'm looking for, I usually end up disappointed for various reasons. Perhaps the rose isn't vigorous or grown on its own roots; or is covered in black spot the moment a rain cloud even thinks about forming; or it doesn't perform well in my hot summer climate; or the thrips feast on it for breakfast, lunch and dinner; or petals shatter the moment it's cut.

Right rose, Right place, Right time.

Don't get me wrong! I LOVE collecting old roses. They are truly gems in my home garden, but many just don't pass the commercial production requirements as cut flowers on my farm. So, my "3R" mantra—Right rose, Right place, Right time—is one I stick to now to save myself future disappointment. You too, can tune into your environment and be observant throughout all 12 months of the year, which will help you to understand how rose varieties perform in your region, based on your growing goals. Learn to let go when the "perfect rose" you wanted really may not be so perfect for your region or climate. In the next chapter, I will introduce three rose-grower "personas" and I hope you see yourself in at least one, as this will help guide you on the path towards your rose-growing goals.

There are a LOT of great rose-gardening texts and reference books available, and I've listed some of my favorites in the reference section at the end of this book. *Growing Wonder* is specifically focused on my wheelhouse: growing garden roses for cut flowers and floral design. As you go through each chapter, you'll learn my tried-and-true practices that I use every day on my farm to grow cut garden roses for professional floral designers.

My goal in writing this book is to shine a light on the best roses for you and your garden or farm, and to give you the confidence to allow space in your life for roses. Keep my positive affirmations in mind as you embark on your rose-growing journey, expand your cutting garden, or join me in the ranks as a flower farmer. If you follow my guidance, you'll discover that the rose is a relatively low-maintenance plant that will reward your attention and care with incomparable flower forms, gorgeous petal palettes, intoxicating scents, and distinctive character in the garden.

'Angel Face', a unique and highly fragrant deep lavender rose, which bears clusters of ruffled, semi-double blooms.

SEVEN ROSE REALITIES

After teaching and coaching hundreds of students about the science and art of growing garden roses, I've probably answered every question imaginable! One thing is for certain: there are plenty of myths and anxieties surrounding the irresistible Rosa genus. I want to debunk misperceptions you've often heard and share my seven "rose realities" to help you get started.

By the conclusion of *Growing Wonder*, if you adopt my mindset outlined in the following chapters, you will discover your own passion for rose growing. I'll give you proven, step-by-step techniques to guide you, whether you have 10 minutes or 10 hours to spend with your roses each week.

A vase of some of my favorite yellow roses includes 'Graham Thomas', 'Teasing Georgia', 'Charlotte', and 'Golden Celebration'.

PHOTOGRAPHY BY JILL CARMEL

You can select great disease-resistant roses.

Make it easy on yourself. Seriously. Select roses for your garden that are newer varieties. When you choose modern introductions that are bred to be more disease-resistant, you'll find roses are relatively easy to care for, and they'll experience minimal diseases (which means fewer interventions from you). Black spot will become a mere annoyance rather than an all-out war. Thank you, modern breeders and scientists, for making it easier on all of us to enjoy roses.

MORE OF MY FAVORITE MODERN ROSE SELECTIONS IN CHAPTER 3.

You can prune like a pro to produce healthy and prolific rose plants.

This is the one task that many rose gardeners and farmers like to let slide, and really, I don't blame them. Who wants to go out in the cold weather and trim up a plant that looks dead in the winter or come home after a long day at work to clip spent blooms?

In the winter season, my inbox is flooded with questions about how to take on the daunting task of pruning thorny, naked, rose canes. I'd venture to say it's the task that most rose gardeners are the most apprehensive about tackling. I'm here to quell your anxiety and give you my step-by-step plan to have you pruning like a pro. Whether it's winter dormant pruning to prepare your roses for the next season or regular deadheading of blooms, in my opinion, pruning is the most important task you can perform to keep your roses healthy and blooming throughout the season.

CATCH MY EASY PRUNING GUIDANCE IN CHAPTER 5.

You can grow roses with sustainable methods.

Roses, just like people, prefer different growing conditions to thrive. My husband would prefer to live in a house that is a frigid 60 degrees F all day, but I like it at a balmy 75 degrees F or I'm reaching for an overcoat. I don't grow pineapples because I don't live in a warm, humid, tropical climate. I know I would be wildly unsuccessful, so I leave pineapple crops to the farmers in Hawaii. The corollary with roses is obvious: Don't grow roses bred for cold climates if your area has 100-degree summers, and don't plant heat-loving roses in the freezing Northeast region.

When it comes to sustainability and growing with minimal inputs and interventions, the key to success is to select varieties that work well in your farm or garden's microclimate. While "sustainable" is often a buzzword that is thrown around without a clear definition of what it actually means in farming and gardening, when it comes to roses, sustainability to me means selecting the best roses for your conditions from the start. If you do your homework and choose wisely, your roses will require fewer inputs and grow well in their natural environment.

LEARN MORE ABOUT SELECTING & GROWING FOR YOUR CLIMATE AND LOCALE IN CHAPTER 4.

You can grow roses in containers.

Any rose can be planted in a pot. Many of my favorite roses are on display in pots and containers of all sizes here on the farm. They add beautiful décor and elegance to the landscape around our home.

Pots are a great way to test out a location if you're not sure a rose will perform well.

Container roses are the best way for beginners to get started with roses, because it takes little commitment to try out, or "trial," a rose that captures your interest.

Containers are also an ideal option for people in apartments, condos, rentals or other locations. You may not have a yard to grow in, but you can enjoy a beautiful garden rose on a balcony or patio.

IN CHAPTER 4, I'LL SHOW YOU HOW CONTAINERS ARE THE MOST ACCESSIBLE WAY TO GROW ROSES.

You can harvest roses for a longer vase life.

Roses like to perform, and like any good performer, when the show's over the make-up comes off and they dash out the stage door. Roses are at their peak performance when they have beautiful open buds and petals unfurled, with pollen and stamens all aglow. Yet, if you clip the flower when it's at full bloom, you've already missed the show. You'll put it in a vase and the petals fall like false eyelashes on the dressing table. Harvesting when the rose bud is tight is ideal. It's like waiting for the show to start, and this method ensures maximum vase life. You won't miss the performance and you'll likely enjoy an encore.

FOLLOW MY INSTRUCTIONS FOR HARVEST AND POST-HARVEST CARE IN CHAPTER 6. ENJOY FRESH-CUT, LONG-LASTING BLOOMS ON YOUR KITCHEN TABLE ALL SEASON LONG.

You can stock your own rose "tool kit" to grow healthy plants.

We've all been there each spring, walking out to check on the amazing explosion of the first flush of roses only to find some sneaky insect has desecrated our beloved blooms. You followed all the tips and tricks to equip your rose plants for success—including proper dormant pruning, applying an organic dormant spray, early-season fertilizing, and pest management. But somehow, Mother Nature appears to be winning.

You feel defeated. I'm here to tell you, you're not doing anything wrong, and you're not alone in your frustration. Our best intentions and care are often thwarted by nature. Even after 20 years of farming, I still go on my morning walk in the field and curse whatever disease or "pest du jour" is currently on attack.

What I can tell you is that I don't let it bother me as much as it did in my early years. I've learned to accept my natural growing environment, including the diseases, insects, and climate challenges, and I now embrace the imperfections and appreciate the natural beauty of each lovely bloom. I know I have my well-stocked tool kit of rose-care strategies to take on any challenge. Whatever nature throws at you your rose tool kit will be well stocked and ready to take on the challenge.

LEARN MORE ABOUT SELECTING & GROWING FOR YOUR CLIMATE & LOCALE IN CHAPTER 5.

You can grow roses in the shade.

I'm not going to lie; roses really are sun-loving plants. To perform their best and put on a stunning show of blooms, they prefer least six-to-eight hours of direct sunlight a day, but they can hold their own in partial shade.

In fact, during the summer months in hot dry climates, like on my farm, garden roses actually like to have a little afternoon shade to keep them cool.

So, if you have a shady alcove or East-facing location with morning sun in your yard, I encourage you to give roses a try there. 'Eden', a climber, or the classic 'Iceberg', are two shade-tolerant roses that I recommend.

YOU'LL FIND A FEW MORE SHADE-LOVING VARIETIES IN CHAPTERS 3 AND 4.

PHOTOGRAPHY BY JILL CARMEL

My arms are filled with an alluring spectrum of just-picked soft and bright pastel roses during a sunset harvest.

THE ROSE PRIMER

Growing Wonder is intended as a year-round guide to growing garden roses for cut flowers, with the specific goal of describing how to grow, tend, and enjoy your garden roses in every season. From the first big display of blooms in the spring, to the naked canes on display in the dormant season, it's useful to familiarize yourself with the stages of a rose plant's lifecycle.

Every year, I fall in love all over again with each beautiful stage of the rose's life cycle. The anticipation of bud break has me watching the plants like an anxious new parent. That joy of bud break is quickly followed by the period of "hurry-up-and-wait," as I watch the calendar while the rose leafs out and its buds form. Then, like a kid on Christmas morning, I rush out to clip the first bloom of the season and put it in a vase on my desk.

> That joy of bud break is quickly followed by the period of "hurry-up-and-wait," as I watch the calendar while the rose leafs out and its buds form. Then, like a kid, I rush out to clip the first bloom and put it in a vase on my desk.

As you establish a rose garden and, if you're like me, continue to expand your collection of garden roses, you will notice how each variety performs through the seasons. Like personalities of people, there is a bit of both nature and nurture in each rose. Some of the traits are in the plant's genetic makeup, and will remain constant throughout its life. You'll notice, though, that some phases of a plant's life cycle may be influenced by cultural conditions, unexpected climate swings, or variables beyond your control.

Some roses need a little more TLC and some, as they say, grow like weeds. Be assured that your rose will continue to progress through its annual life cycle nonetheless! Bloom colors, petal counts, and foliage color will change too—and with each seasonal transition, there is something new and beautiful to admire in your rose collection. Enjoy the unique qualities of your roses throughout the year, and you'll find yourself marking the seasons with confidence.

LIFE CYCLE OF A ROSE

1 DORMANCY
Time for the rose to rest.

2 BUD BREAK
The buds begin to swell and push out of the branch.

3 LEAFING OUT
The newly formed buds push through and unfurl as leaf sets appear.

4 BUDS FORM
Leaves grow an entire stem; rose buds form at the top.

5 FULL BLOOM
Repeats many times throughout the season.

7 ENTERING DORMANCY
Leaves turn yellow and may drop.

6 ROSE HIPS FORM
The fruit of the rose plant, hips form after the flower is pollinated and ripen in late summer through autumn.

GLOSSARY

BLOSSOM: A rose bud that has opened into a flower.

BUD: An immature flower enclosed in a calyx.

BUD UNION: Where the top-stock rose variety and bottom-stock rose variety are fused together on a grafted rose. It's a round bulge from where the canes will grow (There is no bud union on an own-root rose).

CALYX: The cover of the rose bud that is comprised of five sepals.

LATERAL CANE: A secondary cane originating from the main cane. Also often referred to as a side cane or shoot.

LEAFLET: A cluster of leaves together in groups of three, five, seven or more.

MAIN CANE: The main stems of the plant from which lateral canes grow.

ROOTSTOCK: The rose variety on which another variety is budded or grafted on a grafted rose (The roots of an own-root rose will be the same variety as the top stock or canes).

SEPAL: One of five parts of the calyx. They surround the rose bud providing protection until it blooms. The sepals will reflex as the bud opens to a blossom.

ILLUSTRATION BY JENNY MOORE-DIAZ

2

IN THIS CHAPTER

THREE ROSE
PERSONALITIES

WEEKEND
WARRIOR

EVERYDAY
GARDENER

ASPIRING
ROSARIAN

YOUR ROSE PERSONALITY

I wrote *Growing Wonder* to help rose-lovers of all levels turn dreams of having a rose-cutting garden of their own into a reality. Whether I share rose-growing advice with students virtually or teach lessons in person on the farm, it's important to know my first wish for you: Enjoy the journey.

Your success with roses really depends on the level of time you have to spend with them. The good news is that roses respond positively to whatever attention you give to them. They are very forgiving plants. As I tell the members of Menagerie Academy, there are times when you are that helicopter parent of the firstborn child, and other times when you'll just go with the flow and let nature take its course. There are times when you channel Julia Child crafting the perfect dinner party menu, and other times when picking up a rotisserie chicken at the grocery store will have to suffice. In other words, there are many ways to grow a beautiful rose from hyper-focused to more hands off. Whatever time you have, growing roses should be a rewarding and joyful experience!

As a person who has over-ordered tomato seed packets while browsing seed catalogs each winter—only to discover I have bitten off more than I could chew come springtime—take my advice, and make a plan to avoid feeling overwhelmed. Start a wish list, spreadsheet, or garden journal to document the roses you want to grow and note their attributes. When it's all in one place, you can more realistically ask yourself, "what's achievable for me?"

You may already have a list of rose variety favorites you've seen in public gardens, on social media or in magazines. I encourage you to do some research about the requirements of each variety, and whether you can provide what's needed. For example, if you want repeat blooming roses (and who doesn't?), remove those varieties from your wish list that only flower

Rose specimens labeled by variety are ready for a photography session. Menagerie Farm & Flower grows more than 175 different rose types.

PHOTOGRAPHY BY JILL CARMEL

ROSE PERSONALITIES

once each season. If you want low-maintenance roses, look for newer hybrids that have been bred to resist diseases. If you want long, straight stems, read reviews about the health, habit and form of each variety. The Rose Gallery in Chapter 3 is an excellent place to begin.

As your garden begins to mature, be sure to jot down notes about each rose you grow. It's so helpful to look back at previous years to remember the stages in the life of each rose, and note how it bloomed, if it has pest or disease issues, or how it responded to changes in climate and other elements throughout the seasons. As the years go by, you will refine your rose collection as it becomes a beautiful, thriving display of wonderment that's perfect just for YOU!

> As your garden begins to mature, be sure to jot down notes about each rose you grow. It's so helpful to look back at previous years to remember the stages in the life of each rose, and note how it bloomed.

As I guide you through the chapters of *Growing Wonder*, I'll use the following three personality types on the facing page. So be sure to look for the corresponding icon to signal the personality that fits you best, and check out my tips for each type.

These are personality types drawn from the characteristics of my students, and the amount of time they must devote to their roses. How do you fall into this matrix? Remember, depending on the calendar, week-to-week, and season-to-season, you may move between these personas; I know I do.

As the seasons of life change, our available time does too. When I was a new mom with a baby, I only had a few minutes each week to care for the roses outside my kitchen window. Now, during the busy Mother's Day flower rush here at Menagerie Farm & Flower, I am swamped with farm work, and the roses in my garden take a backseat for a few weeks. Don't be defeated if you must scale back or curtail your gardening activities from time-to-time or year-to-year. Give yourself some grace; the roses will be waiting for you when your schedule opens.

WEEKEND WARRIOR

1–2 HOURS PER WEEK

I want the most bang for my buck. I want to achieve a big impact with a small investment of time. Give me the essential tools to enjoy success with just a few beauties!

THREE SUCCESS TIPS

1. With a limited schedule, it's important to commit to spending time with your roses. Set a regular "rose appointment" on your calendar and don't skip it.

2. Select easy-care, disease-resistant and low-maintenance varieties. Leave the roses that are divas for the Aspiring Rosarians to tackle.

3. If you can squeeze in only one care activity each week, make it deadheading. Deadheading regularly ensures you will have beautiful repeat blooms and cut-flower stems all season long.

THREE ROSES TO TRY

'Iceberg'
'Princesse Charlene de Monaco'
'Mother of Pearl'

EVERYDAY GARDENER

3–5 HOURS PER WEEK

I love roses as much as every other flower, but I'd like to learn how to successfully integrate roses throughout the landscape to complement other plantings. Help me become proficient with cut garden roses so I can grow them more confidently.

THREE SUCCESS TIPS

1. The best defense is a good offense. Observe your roses every week for the presence of insects, diseases, and weeds. At the first sign of issues, start control measures to ensure your roses are not overtaken by any of these unwanted invasions.

2. Get a soil test annually. See what your soil or growing media needs, and let it guide your fertilizer and amendment applications throughout the season.

3. Don't skip winter dormant pruning or winter rose care. It's the single, most-important step you can do to prepare for a successful rose-growing season.

THREE ROSES TO TRY

'Bolero'
'Crown Princess Margareta'
'Francis Meilland'

ASPIRING ROSARIAN

2+ HOURS A DAY

I'm all-in for all roses, all the time. I'm passionate about growing all types, and want to learn as much as possible. I'd like a dedicated rose garden where I can harvest roses all season long, or add roses to my small cut-flower farm.

THREE SUCCESS TIPS

1. Keep focused. Set yearly and long-term goals for what you want to achieve. How many varieties do you want to grow? How many stems do you want to harvest? Where and what are you going to use cut flowers for? By setting goals you can make sure not to over-commit to many roses with too little time.

2. Rise and shine. Get out there and harvest in the early morning when it's cool and the roses have tight buds to give your cut stems the longest life.

3. Don't be afraid to remove roses from your collection. Disease-ridden, insect-damaged, small blooms, low-vigor, and short, weak stems are not your friend. They will suck your valuable time that you could be spending on better-performing varieties. Bid those poor achievers a fond farewell.

THREE ROSES TO TRY

'Wollerton Old Hall'
'Honey Dijon'
'Yves Piaget'

3

IN THIS CHAPTER

WHITE
& CREAM

BLUSH

LIGHT PINK

DEEP PINK

BURGUNDY & WINE

LAVENDER
& PURPLE

CRIMSON & RED

GOLDEN & BUTTER

PEACH & COPPER

MULTI & TAUPE

THE ROSE GALLERY

I am enthralled by the incredible character of garden roses—from the color spectrum that ranges from subtle to bold, to the mesmerizing flowers that begin as tiny buds and open to multi-petal blooms throughout the seasons. Garden roses by definition are any roses grown in a garden, and they can transform a space with their style and grace.

There are many classifications of garden roses— from Albas to Rugosas—but in this book, I focused on sharing the varieties and care techniques for the modern roses that I grow for cut-flower production, which include hybrid teas, floribundas, grandifloras, English roses, and climbers.

In this gallery that I've organized by color palette, you will see my favorite garden roses for cutting and floral design. I have grown and evaluated each one on the farm, and these are my top recommendations in each color family. In the pages that follow, you will see: the type of rose; the breeder's name; the color description; the fragrance description; the mature height and spread; the suitable zone; and the average vase life as a cut flower.

Where relevant, I'll note best choices for regions that experience excessive cold or hot temperatures, and a few that perform well in humidity or shadier conditions. In general, these roses will produce flowers that range from three-to-six inches when in full bloom, making them a perfect choice to grow as a cut flower. Use this gallery as a guide to get you started in planning your collection of garden roses for your cut-flower garden or farm.

As you begin to build or expand your collection of roses, there are many different outlets to purchase plants—from brick and mortar nurseries to online suppliers. In the spring season, visitors and workshop students love to walk the fields and see the roses in their element, or peruse our potted

PREVIOUS PAGE + OPPOSITE. A collection of roses reflects the many petal colors, forms, and classifications available for gardeners, flower farmers, and rose aficionados to plant and enjoy.

roses for sale in the nursery. I encourage you to visit a public rose garden, botanical garden, or rose nursery in your area so you can walk through the roses—noting your own wish list as you go. If you're looking for a place to start your shopping, I have listed rose nurseries in the Resource Section to get you started.

While I list the technical features of each rose in this chapter, I can't help but ponder the intangible qualities that may not be as easy to catalog. Garden roses produce a range of feelings such as joy, nostalgia, contentment, and awe, when I see them in my own garden or landscape—not to mention the fragrance that transports me to another time and place. I hope you experience that emotional response as you select and grow them, as well.

TYPE OF ROSE	NUMBER OF DAYS TO BLOOM
Climbing Rose	40-50
English Rose	40-50
Floribunda	45-55
Grandiflora	45-55
Hybrid Tea	40-50

Classifications of Garden Roses

As per the American Rose Society, roses are grouped and classified by the characteristics they possess, and their ancestry.

CLIMBING ROSE: There are three types of climbing roses. The first are large-flowered climbers with tall, stiff canes and blooms in both single- and cluster-flowered canes, which are generally repeat bloomers. Next, are the rambler-type climbers, which have flexible canes that require support that most often flower with small blooms in clusters, usually blooming once per year.

For cut-flower production, all of the roses I use fall into the third category. These are sports or mutations of hybrid teas, floribundas, and others, which resemble their bush counterparts except for their climbing growth habit. They grow taller and climb higher than their regular siblings. Climbing roses are basically roses from other classifications that have a very vigorous growth habit.

ENGLISH ROSE: English Roses—often called David Austin Roses—are technically a group of roses, and not their own class of roses. They were introduced in 1969 by the English-Rose hybridizer David Austin, who bred rose varieties that combined elements of both old roses (roses introduced before 1867) and modern roses (hybrid teas, floribundas and grandifloras), which included cupped-shaped blooms, strong fragrance, and repeat-flowering blooms. Technically, English Roses belong to the shrub class; however, most rose growers consider them a class all their own.

> It helps tremendously to see a mature plant before deciding which roses to take home with you; try to visit your nearest local independent nursery.

FLORIBUNDA: Selected, as the name suggests, for their ability to produce an abundance of blooms in a cluster. The floribunda is a cross between a polyantha and a tea rose. As a class, they are hardier, easier to care for, and more reliable in wet weather than their hybrid tea counterparts. They are also generally shorter in stature than hybrid teas, reaching only three-to-four feet tall. This class can provide long-lasting, year-round garden displays.

GRANDIFLORA: As truly grand blooms, the grandiflora rose is a cross between a floribunda and a hybrid tea. With clustered blooms on long stems, they possess the best qualities of the floribunda and hybrid tea. 'Queen Elizabeth', a tall pink blossom grower reaching six-to-seven feet tall, was the first grandiflora rose to be developed. It was a favorite rose of both my mother and grandmother, and the original plants from their collection still grow on the farm today.

HYBRID TEA: The era of modern roses was established in 1867, with the introduction of the first hybrid tea, 'La France' by the French breeder, Guillot. It was a unique rose that possessed qualities not seen before, with the growth habit of a hybrid perpetual, paired with the elegantly shaped buds and free-flowering character of a tea rose. Today, hybrid tea roses have high-centered and pointed buds, with long stems reaching upwards of six feet and have more than 30-50 petals. While heralded in exhibition circles for one stem to one bloom, hybrid teas can produce clustered blooms if you don't disbud them. Hybrid tea roses produce long, straight-stemmed roses with exceptional vase life for cut flowers.

white & CREAM

You can never go wrong with a classic white rose. Regal in elegance and stature, it is beloved for its simplicity in the garden, and as a cut flower that's always in style for weddings.

'CROCUS ROSE'

AT TOP. A lovely light peach-to-cream shade, this David Austin rose produces layers and layers of petals and steals our heart with every glance. Dreamy rosette blooms open, as this rose turns from soft yellow-to-peach and cream. It's a true chameleon and a must-have for any rose garden.

Type: English Rose (Austin)
Color: Light peach-to-cream
Fragrance: Mild
Height: 4 feet
Spread: 4 feet
Suitable Zones: 5-11
Vase Life: 3-5 days
Characteristics: Heat-tolerant

'WINCHESTER CATHEDRAL'

AT BOTTOM. This beautiful David Austin rose is an early-flowering rose that displays delicate clouds of white rose clusters. The white version of the popular 'Mary Rose', it's one of my most popular selling roses for wedding and event design.

Type: English Rose (Austin)
Color: White
Fragrance: Medium
Height: 4 feet
Spread: 4 feet
Suitable Zones: 4-10
Vase Life: 3-5 days
Characteristics: Cold-tolerant

'TRANQUILLITY'

This is an almost thornless rose with a bright cheery disposition. Its cup-shaped blooms will flush from summer into fall, with nice, long stems, and lush glossy-green foliage.

Type: English Rose (Austin)
Color: Cream
Fragrance: Mild
Height: 4 feet
Spread: 4 feet
Suitable Zones: 5-11
Vase Life: 3-7 days
Characteristics: Heat-tolerant

'EASY SPIRIT'

A creamy rose that is a standout in the field. A beautiful white hue with the most ruffled petals. Amazing disease resistance and lovely dark green foliage.

Type: Floribunda (Carruth)
Color: White-cream
Fragrance: Medium
Height: 5-6 feet
Spread: 3-4 feet
Suitable Zones: 5-10
Vase Life: 3-5 days
Characteristics: Disease-resistant, and heat-tolerant

'BOLERO'

NOT PICTURED. Elegant clusters of white blooms adorn this vigorous, disease-resistant plant that is one of the most fragrant roses I grow. It's a beautiful cut rose for wedding bouquets and event design, and its compact size also makes it ideal for use in containers or pots.

Type: Floribunda (Meilland)
Color: White
Fragrance: Strong
Height: 2-3 feet
Spread: 2-3 feet
Suitable Zones: 5-9
Vase Life: 3-5 days
Characteristics: Cold-tolerant, fragrant, and disease-resistant

'ICEBERG'

A well-loved and iconic white rose with beautiful, prolific blooms. This Kordes-bred classic is a must-have for any garden. Considered perfect for creating a beautiful hedge, it is one of the top 10 roses in the world.

Type: Floribunda (Kordes)
Color: White
Fragrance: Mild
Height: 3-5 feet
Spread: 3 feet
Suitable Zones: 4-10
Vase Life: 3-4 days
Characteristics: Cold-tolerant, disease-resistant, and heat-tolerant

white & CREAM

'FRENCH LACE'

AT TOP. This is my favorite cream rose with beautiful, small blooms that change from ivory-to-light apricot with the seasons. Everybody falls in love at first sight with this compact floribunda.

Type: Floribunda (Warriner)
Color: Ivory-to-light peach
Fragrance: Medium
Height: 3 feet
Spread: 3 feet
Suitable Zones: 5-11
Vase Life: 3-7 days
Characteristics: Cold-tolerant, and heat-tolerant

'WOLLERTON OLD HALL'

AT BOTTOM. As a soft and inviting David Austin climbing rose, this rose is named for the famed English garden. Plump, round buds start as apricot and open to beautiful delicate cream-colored blooms, which are often mistaken for a peony. These prolific plants greet me daily as they line the rose bed outside of my farm office window.

Type: English Rose/Climber (Austin)
Color: Cream-to-blush
Fragrance: Strong
Height: 5 feet
Spread: 3 feet
Suitable Zones: 5-11
Vase Life: 3-7 days
Characteristics: Cold-tolerant, fragrant, and heat-tolerant

blush

From the palest pink to the creamiest peach, blush is a color family that transcends the spectrum. Use it as a blending rose in floral designs to bridge colors in a rainbow of shades. Perfect for a "Blush and Bashful" wedding—as seen in the iconic movie *Steel Magnolias*—or in a simple vase on a bedside table.

'MOONSTONE'

This rose is a classic, upright hybrid tea with big, beautiful, blush blooms that reveal white petals when opened. It has a soft, mild fragrance and is wonderful for cutting with strong straight stems.

Type: Hybrid Tea (Carruth)
Color: Blush
Fragrance: Mild
Height: 4-6 feet
Spread: 2-3 feet
Suitable Zones: 7-10
Vase Life: 3-10 days
Characteristics:
Heat-tolerant

'FRANCIS MEILLAND'

A cream-colored rose named for an iconic rose breeder. It's very fragrant with peach-to-blush tones that give way to soft white as it opens. This rose checks all the boxes as a fantastic cut flower.

Type: Hybrid Tea (Meilland)
Color: Cream-to-blush
Fragrance: Strong
Height: 6-7 feet
Spread: 3 feet
Suitable Zones: 5-11
Vase Life: 3-10 days
Characteristics:
Cold-tolerant, disease-resistant, fragrant, and heat-tolerant

'EARTH ANGEL'

A blush rose with beautiful, romantic, cupped blooms. This Kordes-bred beauty has a classic rose scent and exceptional vigor, and is considered ideal for spring and summer bouquets.

Type: Floribunda (Kordes)
Color: Cream-to-blush
Fragrance: Strong
Height: 6-7 feet
Spread: 3 feet
Suitable Zones: 5-9
Vase Life: 3-7 days
Characteristics:
Cold-tolerant, and disease-resistant

light pink

Like the 1980s ingénue from *Pretty in Pink*, light pink roses come of age in beauty and style. The color is used to represent universal love for oneself and of others. A pink rose forever remains perfect as a token of friendship.

'PRINCESSE CHARLENE DE MONACO'

One of my all-time favorite roses! This ruffled rose is a wonderful cut flower with an unforgettable fragrance and long, straight stems. She is the epitome of style and grace in the garden, and my top pick for a pink-to-blush cut flower.

Type: Hybrid Tea (Meilland)
Color: Soft pink-to-blush
Fragrance: Strong
Height: 6 feet
Spread: 2-3 feet
Suitable Zones: 5-10
Vase Life: 3-10 days
Characteristics:
Cold-tolerant, disease-resistant, fragrant, and heat tolerant

'ELLE'

An exceptional rose that thrives in warmer climates, this rose is a beauty all her own. The petals exhibit a soft-pink blend with hints of orange and yellow emerging as the weather changes with the season. Deep glossy-green foliage makes this a top performer in humid climates.

Type: Hybrid Tea (Meilland)
Color: Pink
Fragrance: Strong
Height: 3-5 feet
Spread: 3 feet
Suitable Zones: 5-11
Vase Life: 3-10 days
Characteristics:
Disease-resistant, fragrant, and heat

'QUEEN OF SWEDEN'

A David Austin rose with long, straight stems that channel a classic traditional upright hybrid tea. Small, cupped blossoms are the perfect fit as an accent in any bouquet.

Type: English Rose (Austin)
Color: Light pink
Fragrance: Medium
Height: 5 feet
Spread: 3 feet
Suitable Zones: 4-11
Vase Life: 3-5 days
Characteristics:
Cold-tolerant, and heat-tolerant

'WISLEY 2008'

With ruffled delicacy and petite blooms, this David Austin embodies the perfect essence of an English-garden rose. Superior disease-resistance and a cup-shaped cluster of petals make this a charmer in the garden.

Type: English Rose (Austin)
Color: Pink
Fragrance: Mild
Height: 5 feet
Spread: 4 feet
Suitable Zones: 5-11
Vase Life: 3-5 days
Characteristics:
Disease-resistant, and heat-tolerant

'ALL DRESSED UP'

A medium-pink rose with long, cutting stems and soft, cupped blooms. This is a classic English-style rose with modern disease-resistance that blooms well in warmer climates.

Type: Grandiflora (Bédard)
Color: Pink
Fragrance: Mild
Height: 6-7 feet
Spread: 3 feet
Suitable Zones: 5-11
Vase Life: 3-10 days
Characteristics:
Disease-resistant, and heat-tolerant

'EVELYN'

Beautiful pink-to-apricot petals adorn this hard-to-find lady named for the iconic perfumers Crabtree & Evelyn, who used it in their range of rose perfumes. This rose is in a class all her own, and a standout from the David Austin breeders. She thrives in warmer climates, and is a glorious choice both in the landscape as well as for cutting and floral design.

Type: English Rose (Austin)
Color: Light pink
Fragrance: Strong
Height: 5 feet
Spread: 3 feet
Suitable Zones: 5-11
Vase Life: 3-7 days
Characteristics:
Heat-tolerant

deep pink

There are no hints of pastel in this popular deep-pink palette range, often associated with the most feminine of roses. In floral design, the roses included here are some of my favorites because they bridge beautifully between paler and darker blooms, and offer tints and shades that move into the berry family of colors. Any of these choices are equally stunning in single-variety floral arrangements.

'DEE-LISH'

AT TOP. With a deep-pink color that is perfect as a cut flower, this rose has excellent disease-resistance and tall stems.

Type: Hybrid tea (Meilland)
Color: Berry
Fragrance: Strong
Height: 6 feet
Spread: 3 feet
Suitable Zones: 5-9
Vase Life: 3-7 days
Characteristics:
Disease-resistant, and fragrant

'SWEET MADEMOISELLE'

AT BOTTOM. With superior disease-resistance and a parentage including 'Graham Thomas' and 'Peace' varieties, this rose has fantastic genetics. Long, straight stems, and exceptional vase life make this stunner a top pick for beginning growers.

Type: Hybrid tea (Meilland)
Color: Pink
Fragrance: Strong
Height: 5 feet
Spread: 3 feet
Suitable Zones: 5-10
Vase Life: 3-10 days
Characteristics:
Disease-resistant, and fragrant

'PRINCESS ALEXANDRA OF KENT'

With a strong tea fragrance, this is just the gal you need for your garden. She displays a familiar David Austin style, including ruffling petals, a beautiful deep-pink color, and resistance to disease.

Type: English Rose (Austin)
Color: Pink
Fragrance: Strong
Height: 4 feet
Spread: 3 feet
Suitable Zones: 4-11
Vase Life: 3-5 days
Characteristics:
Cold-tolerant, and heat-tolerant

'YOUNG LYCIDAS'

This large blooming rose has the most elegant color blending that ranges from deep-pink-to-magenta, with hints of red. Its petals have uniquely ruffled edges, and the bloom produces an old-rose fragrance.

Type: English Rose (Austin)
Color: Deep pink
Fragrance: Strong
Height: 4 feet
Spread: 4 feet
Suitable Zones: 5-11
Vase Life: 3-5 days
Characteristics:
Cold-tolerant, fragrant, and heat-tolerant

'PRINCESS ANNE'

This is a quintessential David Austin rose, with deep-pink blooms that fade to a light red. It is a healthy shrub with strong stems and fragrant clusters.

Type: English Rose (Austin)
Color: Magenta
Fragrance: Medium
Height: 4 feet
Spread: 4.5 feet
Suitable Zones: 4-11
Vase Life: 3-5 days
Characteristics:
Cold-tolerant, disease-resistant, and heat-tolerant

deep pink

'GRANDE DAME'

A glorious fragrance exudes from this bright hybrid tea rose with large, gorgeous blooms. While it is a modern rose, it performs like an old-world classic. It's a perfect addition to your cutting garden with minimal thorns and nice, long stems.

Type: Hybrid tea (Carruth)
Color: Pink
Fragrance: Strong
Height: 5 feet
Spread: 3 feet
Suitable Zones: 5-11
Vase Life: 3-7 days
Characteristics: Heat-tolerant, and fragrant

'JAMES L. AUSTIN'

This stunning deep-pink rose is sure to bring brightness to your garden. It is a versatile shrub with an upright growth habit, as well as a light-to-medium fragrance with hints of blackberry, raspberry, and cherry.

Type: English Rose (Austin)
Color: Deep pink
Fragrance: Medium
Height: 4 feet
Spread: 3 feet
Suitable Zones: 5-11
Vase Life: 3-5 days
Characteristics: Heat-tolerant

'YVES PIAGET'

An old-fashioned variety that evokes memories of your grandmother's garden, this is a striking deep pink rose. With a strong fragrance, this exemplary rose is a favorite of floral designers.

Type: Hybrid tea (Meilland)
Color: Deep pink
Fragrance: Strong
Height: 4 feet
Spread: 3 feet
Suitable Zones: 5-10
Vase Life: 3-7 days
Characteristics: Heat-tolerant, and fragrant

burgundy & WINE

Roses in this group reveal a deep, reddish-brown shade inspired by the wine produced in the Burgundy region of France. These dark-hued blooms are beautiful year-round, but become the stars of the show during the cooler fall months.

'TESS OF THE D'URBERVILLES'

This rich, velvety lady is one of my favorite burgundy roses. It's a David Austin rose that pumps out blooms year-round. Long canes make this rose a wonderful climber, and perfect for cutting nice, long stems.

Type: English Rose/Climber (Austin)
Color: Burgundy
Fragrance: Medium
Height: 4-8 feet
Spread: 4 feet
Suitable Zones: 4-11
Vase Life: 3-5 days
Characteristics:
Cold-tolerant, and heat-tolerant

'MUNSTEAD WOOD'

As a dark and dreamy David Austin rose, this is my go-to for burgundy cut roses. Plump, round buds open to a burgundy stunner with velvet-like petals. This is one of my favorite producers for fall bridal bouquets.

Type: English Rose (Austin)
Color: Burgundy
Fragrance: Strong
Height: 4 feet
Spread: 3 feet
Suitable Zones: 5-10
Vase Life: 3-5 days
Characteristics:
Fragrant

'DARCEY BUSSELL'

This rose is a robust grower and a classic David Austin. It's a lovely rose for fall and winter arrangements with beautiful burgundy-to-crimson ruffled petals and a strong, old-world rose fragrance.

Type: English Rose (Austin)
Color: Burgundy
Fragrance: Strong
Height: 4 feet
Spread: 3 feet
Suitable Zones: 5-11
Vase Life: 3-5 days
Characteristics:
Fragrant

lavender & PURPLE

Lavender, my grandmother's favorite color, transports me back to springtime as a child with her and my grandfather, as we walked through our farm's orchard of French Prunes. This color family will always hold a place in my heart for Mi Abuela.

'CELESTIAL NIGHT'

This dark purple floribunda abounds with exceptional disease-resistance and vigor. I'm totally crazy for this color! With 'Ebb Tide' and 'Grande Dame' as its parents, it has an exceptional pedigree.

Type: Floribunda (Bédard)
Color: Deep-purple to fuchsia
Fragrance: Mild
Height: 6-7 feet
Spread: 3 feet
Suitable Zones: 5-11
Vase Life: 3-10 days
Characteristics: Disease-resistant, and heat-tolerant

'VIOLET'S PRIDE'

As a rose fit for a lady of Downton Abbey, this lavender madame is a continuous bloomer with old English-garden charm. My grandmother would have approved of this new disease-resistant beauty.

Type: Floribunda (Bédard)
Color: Lavender-purple
Fragrance: Strong
Height: 4-5 feet
Spread: 3-4 feet
Suitable Zones: 5-10
Vase Life: 3-10 days
Characteristics: Disease-resistant, fragrant, and heat-tolerant

'INTRIGUE'

With a vibrant deep-purple hue, this rose is perfection. With excellent disease-resistance and an abundance of petals, this rose is a color all its own.

Type: Floribunda (Warriner)
Color: Lavender-deep purple
Fragrance: Strong
Height: 2-5 feet
Spread: 3 feet
Suitable Zones: 5-9
Vase Life: 3-7 days
Characteristics: Disease-resistant

'EBB TIDE'

A smoky-plum in a color family all its own, this rose looks like pure velvet. A strong clove scent, along with beautiful ruffled blooms, makes this one of our most popular roses for cut flowers in the fall and winter months. This tide is a wave we all can get behind.

Type: Floribunda (Carruth)
Color: Deep purple
Fragrance: Strong
Height: 5 feet
Spread: 3 feet
Suitable Zones: 5-10
Vase Life: 3-7 days
Characteristics: Fragrant, and heat-tolerant

'LOVE SONG'

Ruffled lavender buds transition to soft gray petals when open. This rose is a bushy round plant with clusters of large blooms.

Type: Floribunda (Carruth)
Color: Lavender-purple
Fragrance: Mild
Height: 3-4 feet
Spread: 2-3 feet
Suitable Zones: 5-10
Vase Life: 3-7 days
Characteristics: Disease-resistant, and heat-tolerant

'QUEEN OF ELEGANCE'

What happens when you take 'Koko Loco' and 'Life of the Party' and put them together? You get 'Queen of Elegance'. She has the most beautiful fading color like her mom Koko, and is a rose fit for royalty. It's a new addition to the rose community with a unique color, making it a must-have addition to any rose collection.

Type: Floribunda (Carruth)
Color: Deep pink-to-purple
Fragrance: Strong
Height: 5 feet
Spread: 3 feet
Suitable Zones: 5-10
Vase Life: 3-7 days
Characteristics: Disease-resistant, fragrant, and heat-tolerant

crimson & RED

The colors of love and passion, danger, anger, and adventure. A single red rose is an iconic gift to show affection.

'STILETTO'

This rose is a standard hybrid tea rose with long stems and pointed buds. Its striking red-to-magenta tones make it destined to become a classic. This rose is a bit particular about its growing conditions though—preferring warmer winter climates—so I recommend it for gardens in Zones 6 and above.

Type: Hybrid Tea (Meilland)
Color: Classic red
Fragrance: Strong
Height: 5 feet
Spread: 3 feet
Suitable Zones: 6-10
Vase Life: 3-7 days
Characteristics: Disease-resistant, fragrant, and heat-tolerant

'RED TRAVIATA'

With old-fashioned English-style blooms, this is a beautiful rouge-colored rose that is considered perfect for fall and winter floral design.

Type: Hybrid Tea (Meilland)
Color: Red
Fragrance: Mild
Height: 4 feet
Spread: 3 feet
Suitable Zones: 5-9
Vase Life: 3-7 days
Characteristics: Disease-resistant

'SEDONA'

Strong stems and continuous blooms throughout the season make this rose a beautiful garden rose. Its unique petal color has shades of red that fade to a burnt-orange that resembles a sunset.

Type: Hybrid Tea (Zary)
Color: Crimson-red
Fragrance: Strong
Height: 4-5 feet
Spread: 3-4 feet
Suitable Zones: 5-10
Vase Life: 3-7 days
Characteristics: Disease-resistant

'ROUGE ROYAL'
This very large blooming rose is unique with its petal shapes. It opens to a bright, red raspberry color and sweet citrus fragrance.

Type: Hybrid Tea (Meilland)
Color: Crimson-red
Fragrance: Strong
Height: 5 feet
Spread: 3 feet
Suitable Zones: 5-9
Vase Life: 3-7 days
Characteristics:
Disease-resistant, and fragrant

'MR. LINCOLN'
With velvety red petals and dark green foliage, this is a classic long-stemmed red rose ideal for cutting. If you're looking for the quintessential red rose, look no further.

Type: Hybrid Tea (Swim)
Color: Red
Fragrance: Strong
Height: 4-6 feet
Spread: 2-3 feet
Suitable Zones: 5-10
Vase Life: 3-10 days
Characteristics:
Fragrant, and heat-tolerant

'LAVA FLOW'
NOT PICTURED. Rich, deep-red, ruffled clusters adorn this compact bush. It produces an eruption of color and a beautiful rose for red rose lovers. Petite clusters of blooms make this a perfect spray rose for cutting, and it has exceptional vase life.

Type: Floribunda (Kordes)
Color: Crimson-red
Fragrance: Mild
Height: 2-3 feet
Spread: 2-3 feet
Suitable Zones: 5-10
Vase Life: 3-10 days
Characteristics:
Disease-resistant, and heat-tolerant

'HOT COCOA'
With tones of chocolate and burnt-orange, this rose is like a chameleon. Large clusters and fragrant flowers make this a unique rose to add to your collection.

Type: Floribunda (Carruth)
Color: Crimson-red
Fragrance: Moderate
Height: 4-5 feet
Spread: 4 feet
Suitable Zones: 5-10
Vase Life: 3-7 days
Characteristics:
Disease-resistant, and heat-tolerant

golden & BUTTER

Yellow roses are my absolute favorite color of rose! With a bright and cheery disposition, they truly are sunshine on a cloudy day.

'CHARLOTTE'
A lovely yellow shade that will brighten any garden, this David Austin rose performs well in both warm and cool climates. With cupped blossoms, it stays rather compact for an English rose.

Type: English Rose (Austin)
Color: Butter yellow
Fragrance: Strong
Height: 4 feet
Spread: 3 feet
Suitable Zones: 5-11
Vase Life: 3-5 days
Characteristics: Cold-tolerant, fragrant, and heat-tolerant

'GOLDEN CELEBRATION'
A true-to-type David Austin rose, this vibrant yellow rose possesses the vintage English-garden rose traits. With large buds and an upright stance, it's a shining star in any garden setting.

Type: English Rose (Austin)
Color: Golden yellow
Fragrance: Strong
Height: 5 feet
Spread: 4 feet
Suitable Zones: 5-11
Vase Life: 3-7 days
Characteristics: Fragrant, and heat-tolerant

'MICHELANGELO'
This rose is a farm favorite with ruffled yellow petals, a citrus fragrance and glossy green foliage. An absolutely perfect rose for cutting, it will add a splash of brightness to your home or garden.

Type: Hybrid Tea (Warriner)
Color: Yellow
Fragrance: Strong
Height: 5 feet
Spread: 3 feet
Suitable Zones: 5-9
Vase Life: 3-10 days
Characteristics: Cold-tolerant, disease-resistant, fragrant, and heat-tolerant

'TEASING GEORGIA'

A lovely light peach-to-yellow shade that changes with seasonal temperatures, this David Austin rose performs well in both warm and cool climates. Gorgeous long stems make this cut flower a beautiful climber that can be maintained as a shrub if desired.

Type: English Rose/Climber (Austin)
Color: Peach-to-pale yellow
Fragrance: Strong
Height: 12 feet
Spread: 3 feet
Suitable Zones: 5-11
Vase Life: 3-10 days
Characteristics: Cold-tolerant, disease-resistant, fragrant, and heat-tolerant

'MOONLIGHT ROMANTICA'

A vigorous bloomer, this Kordes-bred rose is the perfect shade of golden yellow. This rose hits all of the marks as a superb cut flower, while being very fragrant and disease-resistant too.

Type: Hybrid Tea (Kordes)
Color: Golden yellow
Fragrance: Strong
Height: 6 feet
Spread: 3 feet
Suitable Zones: 5-10
Vase Life: 3-10 days
Characteristics: Cold-tolerant, disease-resistant, fragrant, and heat-tolerant

'GRAHAM THOMAS'

A true-to-type David Austin selection, this vibrant yellow English-garden variety is a classic. With large buds and an upright stance, it's a great addition to your garden landscape. It can be grown as a climber or pruned regularly to a shrub height during the growing season.

Type: English Rose/Climber (Austin)
Color: Golden Yellow
Fragrance: Strong
Height: 12 feet
Spread: 4 feet
Suitable Zones: 5-8
Vase Life: 3-5 days
Characteristics: Fragrant

peach & COPPER

This warm spectrum of colors represents some of the most popular on my farm. They exhibit a blend of shades that includes orange, yellow, and even touches of white. In both floral design and garden landscape, it's a comforting palette that evokes warmth, joy, and happiness.

'CROWN PRINCESS MARGARETA'

As one of my favorite peach roses here at the farm, this David Austin charmer has beautiful rosette-shaped blooms and a pleasant fruity fragrance. With nice, long canes, this rose can be grown as a climber or cut regularly for shape, making a wonderful shrub.

Type: English Rose/Climber (Austin)
Color: Peach
Fragrance: Strong
Height: 4-12 feet
Spread: 4 feet
Suitable Zones: 4-10
Vase Life: 3-5 days
Characteristics: Cold-tolerant, fragrant, and heat-tolerant

'MARILYN MONROE'

Creamy, light-apricot petals are perfect for a rose named for the iconic Marilyn. A vigorous producer of soft roses, especially in warm climates, this rose has exceptional vase life. Be careful though, this lady is very thorny.

Type: Hybrid Tea (Carruth)
Color: Light peach
Fragrance: Mild
Height: 4-6 feet
Spread: 3-4 feet
Suitable Zones: 6-10
Vase Life: 3-10 days
Characteristics: Disease-resistant, and heat-tolerant

'LADY EMMA HAMILTON'

With ruffled delicacy, this David Austin rose is a perfect symbol of an English-garden rose. It has superior disease-resistance, with unique red-hued foliage and a classic cupped bloom. The sweet scent on this lady will blow you away, making it one of my all-time favorite roses.

Type: English Rose (Austin)
Color: Orange-peach
Fragrance: Strong
Height: 4 feet
Spread: 4 feet
Suitable Zones: 5-11
Vase Life: 3-7 days
Characteristics: Disease-resistant, fragrant, and heat-tolerant

'VALENCIA'

Soft peach petals decorate this rose with generously large blooms spanning up to six inches in full bloom. It has a mild scent with good disease-resistance, and luscious, long-stemmed blooms throughout the growing season.

Type: Hybrid Tea (Kordes)
Color: Apricot
Fragrance: Mild
Height: 5 feet
Spread: 4 feet
Suitable Zones: 5-9
Vase Life: 3-7 days
Characteristics: Disease-resistant, and heat-tolerant

'CARDING MILL'

Beautiful pink-to-apricot petals adorn this David Austin garden rose. It is a wonderful repeat bloomer and loves a warmer climate. From a deep coral in cooler weather, to a light peach in the summer sun, it's a glorious choice as a cut flower.

Type: English Rose (Austin)
Color: Apricot, peach and copper
Fragrance: Medium
Height: 4 feet
Spread: 4 feet
Suitable Zones: 5-11
Vase Life: 3-5 days
Characteristics:
Fragrant, and heat-tolerant

'MOTHER OF PEARL'

As a consistent bloomer throughout the season, this rose adds a simple elegance to any floral arrangement with petals that are almost iridescent. It is an exceptional performer in cooler climates, and is resistant to blackspot in humid locales.

Type: Grandiflora (Meilland)
Color: Peach
Fragrance: Mild
Height: 4-5 feet
Spread: 4 feet
Suitable Zones: 5-9
Vase Life: 3-7 days
Characteristics:
Cold-tolerant, and disease-resistant

multi & TAUPE

Like chameleons changing colors and tones with the seasons, the roses in this family are the most unique in my collection. From mauve to glossy purple, and a kaleidoscope of colors in between, the roses in this group always steal the show.

'STATE OF GRACE'

AT TOP. This is a beautiful multicolored rose with shades of pink-to-gold and copper. She'll add elegance and old-world style and grace to any garden.

Type: Grandiflora (Bédard)
Color: Apricot-to-gold
Fragrance: Medium
Height: 5 feet
Spread: 3 feet
Suitable Zones: 5-10
Vase Life: 3-7 days
Characteristics:
Disease-resistant, and heat-tolerant

'HONEY DIJON'

AT BOTTOM. As the darling of wedding and event floral designers, this rose almost needs no introduction. Its unique mustard color with pink-streaked tips puts it in a class all by itself. Its parents are two exceptional roses: 'Stainless Steel' and 'Singing in the Rain'. It is one of the most popular cut garden roses here on the farm.

Type: Hybrid Tea (Sproul)
Color: Honey mustard
Fragrance: Medium
Height: 4 feet
Spread: 3 feet
Suitable Zones: 5-10
Vase Life: 3-10 days
Characteristics:
Disease-resistant, and heat-tolerant

'KOKO LOCO'
She almost needs no introduction—this crazy rose goes loco with shades of lavender-to-taupe while blooming. This rose is the sweetheart of floral designers and trendsetters.

Type: Floribunda (Bédard)
Color: Multicolor-taupe-lavender
Fragrance: Mild
Height: 3-4 feet
Spread: 3-4 feet
Suitable Zones: 6-9
Vase Life: 3-7 days
Characteristics: Disease-resistant

'DISTANT DRUMS'
As a beauty of an ombre rose, this is a favorite of floral designers and gardeners alike. It produces flushes prolifically throughout the season, and is one of my all-time favorite roses. Everyone who meets this beauty falls in love with it.

Type: Floribunda (Buck)
Color: Multicolor-taupe-lavender
Fragrance: Medium
Height: 3-4 feet
Spread: 3 feet
Suitable Zones: 5-9
Vase Life: 3-10 days
Characteristics:
Disease-resistant, and heat-tolerant

'WORLD WAR II MEMORIAL'
This is a pale-lavender rose that looks almost iridescent when hit by the morning sun. It is very fragrant with long, straight stems that are perfect for cutting.

Type: Hybrid Tea (Weeks)
Color: Lavender-to-light pink
Fragrance: Strong
Height: 4 feet
Spread: 2 feet
Suitable Zones: 6-10
Vase Life: 3-7 days
Characteristics:
Fragrant, and heat-tolerant

WHERE, WHEN & HOW TO
PLANT ROSES

There is something irresistible about a rose's mystique. We are drawn in and our senses want more from silky, soft petals, unforgettable fragrance, and petal colors only Mother Nature could create. The same allure that entices us to desire this special flower also often leaves even seasoned green thumbs in a panic when faced with the myriad of choices to make when planting a new rose.

It's easy for experienced and newbie gardeners alike to feel apprehension about achieving a healthy start when it comes to rose selection, placement, and planting.

In this chapter, I'll to show you how effortless they really are to grow! I know it can be easy to feel overwhelmed with all of the choices between bare root or potted; own root or grafted; optimal planting times; and ideal placement, not to mention considerations about your growing region! It seems like a long checklist to review, and you feel overwhelmed about what's best for you and your garden. Does this sound familiar?

Honestly, while roses do have that unmistakable air of mystery, they are much simpler than their aura and reputation convey. Like most plants, garden roses just need four basic things: sunshine, water, well-drained soil, and the right spot to put down roots and bloom. I'm going to review and define rose-growing lingo as I break down your choices into simple steps, and your roses will flourish in no time.

NOTE: This book is geared to rose gardeners and growers in the Northern Hemisphere; all zone references are from the USDA Plant Hardiness Zone Map.

4

IN THIS CHAPTER

SELECTING
A TYPE OF ROSE

WHERE AND WHEN
TO PLANT

TIME TO PLANT

PLANTING
A BARE ROOT ROSE

PLANTING
A POTTED ROSE

TRANSPLANTING
A ROSE

Early in the season, depending on your zone, it's time to select and prepare bare root roses for planting.

SELECTING A TYPE OF ROSE

Who doesn't love a good day of retail therapy at the garden center? To me, plant shopping makes me feel like a kid in the proverbial "flower" candy store. Whether you're surfing the internet for bare root roses during the cold days of winter as you dream of blooms to come, or wandering the rows at your local garden center when the first shipment of potted roses arrives in early spring, the excitement of adding roses to your collection brings a rush of adrenaline.

However, before scheduling your shopping trip, check out Chapter 3 for some of my favorite cutting roses to grow.

How to Select Roses for Your Climate

Look at the care card, information guide, catalog or internet product listing that accompanies the rose plant. It will have important information on the height, spread, and hardiness zone. Choose a rose that is rated hardy for your growing region. Some varieties are bred to withstand cold or freezing winter temperatures, while others are better suited in hot, arid climates. The care card will always have information on the color and other special growing characteristics, such as if a rose is resistant to blackspot and disease.

Purchase from a reputable rose nursery. I have several listed in the Resource Section. A good rose nursery—especially one that is local to your area—will be able to provide suggestions for roses that will be successful in your specific climate.

Connect with your local rose society. Local Rosarians and rose aficionados are a wealth of knowledge about the growing conditions and roses that are stars in their neck of the woods.

Visit the "Help Me Find Roses" website (helpmefind.com). It has a database cataloged with information on more than 44,000 roses, including 160,000 photos. It's a great place to begin an online search for all-things roses. Join the easy-to-use online message boards to find like-minded rose-lovers in your growing region.

Join a local or online garden club. Garden clubs are a great way to connect with fellow rose enthusiasts to pick their brain, swap stories about which roses are star performers, and which ones are clunkers.

Bare Root vs. Potted and Grafted vs. Own Root

There are two different types of roses you can purchase and plant. Roses in your garden can be started from either bare root plants or potted plants grown in a soil media. Many mail-order and online nurseries offer bare root plants or one-gallon potted plants (sometimes called a "band pot" or "banded roses") that are available for shipping across the country.

Local nurseries, garden centers, and large retail chains offer both bare root roses and three- or five-gallon containers for pickup at their brick-and-mortar locations. Bare root roses are available in the winter and early spring (usually during your region's dormant season), while container roses are available beginning in the early spring through the fall.

So, what should you choose: a bare root rose, band pot, or potted rose? Honestly, all options will give you a great start growing roses; however, as a cut-flower farmer, I prefer to start my roses from bare root plants or three- to five-gallon potted roses. Band pots are basically a rooted cutting of a rose that just isn't quite as far along in its lifespan and requires more work, attention, and time to produce a full-sized rose with cuttable stems.

BARE ROOT ROSES	**POTTED ROSES**
Available seasonally in the winter and early spring	Available almost year-round depending on growing region
Half the cost of a five-gallon potted rose	Double the price of a bare root rose
Wide availability of varieties from online and mail-order nurseries	More limited selection of varieties
Rare and hard-to-find varieties available from online and mail-order nurseries	Selection of varieties intended to perform well in your local climate
Must be planted in the dormant season before warmer spring weather arrives	Can be planted virtually year-round if your ground isn't frozen
Roots faster and gets established in its location more quickly	Can become root-bound if plant remains in the pot for too long
Canes present at planting need time to grow before producing foliage	Provides an instant garden with foliage, blooms, and fragrance

I'm the type of gal who wants more instant gratification with stems I can cut and sell as soon as possible, and I find most of my nursery customers feel the same. They want to be able to plant a rose and have a healthy, sizeable, thriving plant with cuttable stems within the first year. For this reason, I save my banded pot collection for an area of my garden that needs more patience, or those varieties I'm willing to grow with patience, as they need a few years to develop into a plant with long, cuttable stems. If you're just starting your rose-cutting garden or adding roses to your flower farm, stick to bare root roses and larger-sized potted roses for a quicker return on your investment.

Bare Root 101

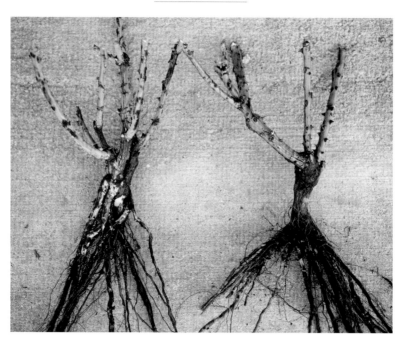

ABOVE. An own-root rose and a grafted rose.
OPPOSITE PAGE. Come spring, the nursery at Menagerie Farm & Flower is stocked with a healthy selection of potted roses, ready for planting.

GRAFTED ROSES: Grafted roses, also called budded roses, are plants that have a top-stock variety grafted to a bottom stock from a different variety. The shoots (aka canes) that grow above the ground are a different variety than the roots that grow below ground. The top stock is "grafted" or fused to the bottom stock, so they grow together as one plant. It's like taking the top and bottom of two different

WHICH SHOULD YOU CHOOSE?

YOUR ROSE PERSONALITY

WEEKEND WARRIOR

Time is valuable, so make it easy on yourself! Visit a local nursery to select five-gallon potted roses in the spring. Look for rose plants that are leafed out and blooming. This strategy brings instant color and plenty of blooms to your garden without the wait. Don't mess with bare root roses unless you have a three-day weekend coming up with time to prep and plan for winter planting roses.

EVERYDAY GARDENER

Try a mix of both bare root and five-gallon potted roses. Don't be afraid to try both grafted and own root roses for the greatest selection of varieties. Grade 1 bare roots are easy to establish when you follow my simple planting steps. Shop both online and with local retailers for a diversity of varieties. Shop early in the season for the best selection of both bare root and container-grown roses.

ASPIRING ROSARIAN

Dive right in and make your wish list for bare root roses. Online and catalog nurseries offer lesser-known choices, and often stock hard-to-find varieties that you can't find in big-box chains and local garden centers. You'll be able to score some beautiful gems for your collection at about half the price of five-gallon potted roses.

PHOTOGRAPHY BY JILL CARMEL

broken crayons, taping them up and melting them together to form a new single crayon. Why is this done? The bottom-stock variety makes the rose plant hardier and healthier. Rose growers select specific bottom stock to give the rose plant improved disease-resistance, hardiness, and vigor in adverse weather conditions.

OWN ROOT ROSES: Own root roses are grown from a single-variety cutting that develops its own root system. Unlike grafted roses, both their roots and shoots come from the same variety of plant.

WHICH TO CHOOSE: There is no one type that is better than the other. Both own root and grafted will grow beautiful, hardy roses. So, select any variety you'd like, and you'll be on your way to a beautiful rose garden come spring!

ZONE 3-5 SPECIAL CONSIDERATIONS: You can grow grafted roses successfully in Zones 3-5, but your plants will require more winter protection and care to ensure the top-stock variety stays viable the following year and does not die back when the temperatures drop. If you're new to growing roses in Zones 3-5, and not sure how a grafted rose might perform in your climate, try out one or two for an initial season to see how they fare. If you don't want to take a gamble on grafted roses, begin with own-root varieties and you'll be off to a good start.

Bare Root Grade Selection

The grade indicates the size of the rose when it's harvested from the growing field. Most commercially available American bare root rose plants are grown in fields by the millions in California, Arizona, and Texas. During the fall of their first or second year, they are harvested or dug out of the growing fields. The soil is shaken off of the roots and the plants are graded, watered, and placed in a moist, cold storage environment to replicate the dormant conditions they would have in nature.

This storage technique puts growth on pause, leaving the rose in a state of suspended animation as it waits to be planted in home gardens or potted by production nurseries into containers. These naked plants live on their own stored energy during that brief stint in cold storage. From storage, they are shipped across the country to nurseries and home gardens from December to early May.

A rose grade reflects the number of canes a rose must have, the size of the canes, and where on the plant measurements are taken. Over 50 years ago, the American Association of Nurserymen—in association with the American National Standards

> **DID YOU KNOW?**
>
> Bare root roses purchased by mail-order or online can be planted in any climate regardless of where they are originally grown. So, it's no problem to take a rose grown in California and plant it in your garden in Maine, Michigan, or Florida. The rose may take some time to break dormancy, but once established, it will be laden with blooms.

Institute (ANSI)—developed grading standards for grafted, field-grown garden roses in the United States to standardize the quality of rose sizes sold by all retailers. A Grade 1 rose is the top grade; Grade 1.5 is one step down; and Grade 2 roses are the smallest grade. A Grade 1 grafted rose needs three canes branched no higher than three inches above the bud graft, measuring at least 5/16-inch in diameter. A Grade 1.5 rose has at least two strong canes, while a Grade 2 needs one 5/16-inch cane, and at least one at 1/4-inch. Grade 1 is the most expensive and prices decrease by grade accordingly.

Grafted roses are typically larger in size than own root roses. Roses of the same variety can be labeled Grade 1 while also being physically different sizes. This is because the standardized ANSI grading system is different for grafted rose and own root rose categories. Bare root roses that are imported from Canada or other countries also may be labeled as Grade 1, but look smaller in size than their U.S. counterparts. This is one area of rose growing where size doesn't always matter. What's important is looking for the grade number, whether it's grafted or own root, and U.S.-produced or imported.

I recommend choosing a Grade 1 rose when shopping for plants (either grafted or own root), because it will be the healthiest plant with strong canes, which ensures your rose gets off to a great start. The adventurous gardener and budget-seekers alike can often score some great Grade 1.5 or Grade 2 varieties at discounted prices. A quick trip around the back aisles at your local big-box home improvement center at the end of the season will also yield nice bare roots at clearance prices. In choosing grades, it ultimately depends on your budget and personal preference.

When selecting bare root roses of any grade look for firm, green canes. Avoid mushy, broken or cracked canes, or plants with soft or moldy roots. Dried-out roots and canes are the number-one reason a bare root rose won't survive when planted. Whether it's from an online nursery supplier, or from a local garden center, make sure the roots are supple, and that you keep them moist until you are ready to plant.

WHERE & WHEN TO PLANT

You selected your favorite roses and now what? It's time to choose your planting locations. While garden roses are extremely resilient plants and can be established just about anywhere, they do have conditions that are optimal for success. If you have a less-than-ideal spot in the landscape, don't worry. Many issues can be mitigated with a little pre-planning and TLC.

For example, if your soil has poor drainage, you can plant roses in raised beds or in containers. Or, you can add additional organic matter when planting. If you have dappled shade, you can select varieties that are shade-tolerant. Are you in an area with too much sun or intense heat in the summer months? Plant in a location that gets afternoon shade or erect a shade cloth for cover. Commercial flower farmers in colder regions (Zones 3-5) can even grow garden roses in low tunnels or greenhouses that mimic conditions one to two zones warmer than their zip codes. Whatever the conditions at your farm or garden, I'm sure there's a location perfect for growing beautiful roses.

> There are many ways to incorporate cut-garden roses into your farm or garden, from mixing roses into the existing landscape to planting in pots or containers.

Let the sunshine in: Roses are sun-lovers. Select a location that has six-to-eight hours of sunlight, well-drained soil, and minimal wind. Once you have identified the site, take a soil test. A quick internet search will lead you to a local soil lab, or you can check out my Resource Section for companies who provide mail-in soil test kits that service the entire U.S. Also see, a list of the optimum pH and soil nutritional requirements for growing roses.

If your soil test comes back with flying colors, then you're good to go. If the test results recommend adding additional organic matter, fertilizer, or other amendments, you can do so before planting. If your soil test results were less-than-favorable, consider selecting an alternative location to plant, or plant in a raised bed or container filled with a potting soil mix.

QUICK SOIL-DRAINAGE TEST FOR PLANTING

Roses need water to thrive, but they don't like wet feet or having a puddle at their base. A quick way to determine the water-holding capacity of your soil is to do this simple test:

STEP 1: Dig a 1-1.5-foot-deep hole where you want to plant your rose.

STEP 2: Fill the hole with water.

STEP 3: See how long it takes for the water to drain into the ground.

If it takes longer than eight hours to drain, then you have soil that drains poorly. To remediate poor drainage, you can choose to plant in a raised bed or in a container filled with potting mix. Or you can dig a planting hole in your native soil that's two feet deep or deeper.

Amend the hole with compost or other organic matter—the addition of organic matter will create air space in the soil allowing the water to drain more easily. Once amended, you can try the "dig a hole and add water" test again to see how long water takes to drain. Continue to amend the planting area with more applications of organic matter as necessary.

Now that you have the perfect spot for your roses, it's time to make a planting plan. How do you want your rose garden to appear or feel? Do you envision something modern, elegant, and minimal like a Palm Springs bungalow garden, or do you dream of having an English-style cottage garden? Does something formal appeal to you—complete with a bench at the center of a symmetrical space? Maybe you want a simple cutting garden to supply your vases, or perhaps need taller roses that double as a screening hedge, or desire easy-care roses to line the border of a walkway?

Once you determine your design aesthetic, you can finalize your planting map. (Check out the Resource Section for some of my favorite garden design books for inspiration.) There are many ways to incorporate cut-garden roses into your farm or garden, from mixing roses into the existing landscape to planting in pots or containers. Roses can bring color and texture to a patio garden, or can be just as successfully lined up in row formation for commercial cut-flower production.

Garden roses can be planted two-to-four feet apart depending on the variety, growing zone, and growing goals. They can also be planted farther apart, depending on your preference. For home gardeners and flower farmers alike, I recommend a three-foot plant spacing in any growing zone. While there are exceptions to every rule, I find this is a great distance for most growers who desire a beautiful rose-cutting garden. This spacing pattern hits all marks, and allows you to clip blooms efficiently from your rose plants while maintaining a nice display in the landscape. With this spacing, plants receive good airflow and canes won't get damaged from adjacent plants because of windy conditions. Plus, you'll have space to maneuver, harvest, and care for your cut-rose collection.

> **THINGS TO KEEP IN MIND AS YOU PLAN YOUR ROSE SPACING**
>
> **Make sure to leave room around the rose,** or cluster of roses, so you can walk around them for deadheading, harvesting, and general care. Remember, you're planting a cut-rose garden and one of the most important things you need to do is be able to easily cut stems.
>
> **Check the rose tag before purchase** to note the expected height and spread of a full-grown plant. Varieties that produce a smaller spread can be planted closer together. Likewise, plants that have a larger spread can be spaced further apart.
>
> **Consider your palette.** Do you want a patchwork of colors and sizes, or should you color-block groups of roses for ease of harvest?

If you followed my advice and planted three feet apart, but are now finding this layout isn't working for you, no problem! You can move roses as easily as rearranging furniture in the living room. That's right, during your dormant season, you can dig up any rose and change the spacing or move it to a new location entirely. Trust your grower's instinct and don't be afraid to alter the location or spacing. Your rose babies will survive the replanting. (See my transplanting tips at the end of this chapter.)

If you are in a climate with warm, humid summers, it's better to plant a little farther apart (three-plus feet apart), increasing the airflow to prevent disease

Rose plant spacing takes on an orderly pattern at Menagerie Farm & Flower.

ROSE PLAN AND PREP

YOUR ROSE PERSONALITY

WEEKEND WARRIOR

Dive into the rose-growing game by planting in a pot or raised bed. It's easy to prepare these areas, so you have the perfect soil to begin, with minimal long-term commitment.

If you don't like the rose or location, you can move it or bid it a fond farewell. As your rose garden evolves, you can either leave the rose in its pot or plant it in the ground in a permanent location.

EVERYDAY GARDENER

Set the right foundation for your roses for years to come with proper bed preparation from the start. Take the time to amend your planting area before planting with compost and organic matter.

ASPIRING ROSARIAN

You can pack a lot of roses into a smaller garden space. (If you're like me and want to order everything, this is good news!) Don't be afraid to plant roses with closer spacing, and regularly shape-prune them throughout the year to the size that's ideal.

like black spot (or just better yet, choose disease-resistant roses). If you are in a colder climate or warm, arid region, you can plant roses two feet apart for a nice hedge or border.

If you are a flower farmer looking to grow roses for cut-flower production, roses can be planted closer, but they should be offset in placement. To make the most of your planting space, group roses by variety and/or color for maximum harvesting efficiency.

In short, like most rose practices, there is no perfect plant spacing rule. If you remember my "3R" mantra—Right rose, Right place, Right time—you'll find that through trial and error, and after gaining a few growing seasons under your belt, you'll find the perfect plant spacing for your garden or farm.

Planting Roses in the Shade

Roses really are sun-loving plants. I imagine them happiest on a permanent vacation, seated in a lounge chair, sunning themselves by the pool with a tropical drink and a smile on their face. They require six-to-eight hours of sun per day to really put on a show-stopping display. You can, however, enjoy roses in partly shady areas by revamping your expectations of the rose's performance and vigor.

> Garden roses that are good for cutting usually have desired characteristics like higher petal counts and larger blooms; finding varieties that do well in shade can be challenging.

It's important to remember that roses use the sun through photosynthesis to feed the plant, so less sunlight means the rose can't produce the food it needs to thrive. I receive hundreds of emails each year from gardeners wondering why a rose they have is struggling. Their rose doesn't have noticeable disease or insect issues; it's getting adequate water and care; so what could be the problem?

Usually after learning more details from the rose owner, I conclude their challenge is because the rose is planted in partial, or almost total, shade. When a rose doesn't

> **WHEN IS IT OKAY TO PLANT IN THE SHADE?**
>
> As you'll find throughout this book, you can always find exceptions to just about any growing advice. Rose growers in warmer climates, such as in Zones 9 or 10, may actually PREFER to plant some of their roses in a location that gets late afternoon shade.
>
> Or, during the hot summer months, you can erect a shade cloth over the top of full-sun roses. I fall into this category. I plant in full sun, but like the flexibility of adding a shade cloth to my potted roses in our farm nursery during the hot summer months.

get the energy from the sunlight to perform photosynthesis, you end up with a rose that doesn't rebloom often, with smaller-sized blooms, minimal petal counts, and an overall smaller plant with thinner canes and weaker stems.

As you are scouring rose catalogs and lists, keep this in mind: In general, roses that have fewer petal counts naturally will grow better in shade because they don't need as much energy to produce their blooms. Roses with single or smaller petals will also dry out faster if hit by morning dew, which prevents possible diseases that may creep into a shadier locale.

Since garden roses that are good for cutting usually have desired characteristics like higher petal counts and larger blooms, finding varieties that do well in shade can be challenging for a cut-rose grower. I've done the hard work for you and suggest the following varieties to try. These are great as cut flowers AND will thrive in shadier locations: 'Sally Holmes', 'Iceberg', 'New Dawn', 'Eden', 'Mary Rose', and 'Winchester Cathedral'.

Remember, any rose that is planted in a shady spot will likely have smaller blooms and petal counts than their varietal counterparts planted in full sun. So keep this is mind when you are comparing rose performance in shady versus sunny homes.

If you're unsure whether a rose will perform well in shade, a great test is to plant the rose in a pot or container. Trial that rose for a season or two, and if it performs the way you like in the shade, then you can plant it in the ground at that location. If it doesn't perform as well as you'd like, move it to a location with more sun.

WHERE, WHEN, & HOW

TIME TO PLANT

You've selected the perfect rose and location and now it's time to plant. Roses can be planted virtually year-round in warmer climates, and in the winter-to-late-summer in all climates. You can use the U.S. Department of Agriculture Hardiness Zone map to help you identify your growing zone. Divided into 13 zones that span the country, this map is a great starting point to learn more about the climate where you live. There are also hardiness maps that divide the countries of Canada and Australia, as well as Europe. Wherever you call home, you will likely be able to find a growing zone map for your plot of land. You can find hardiness maps at planthardiness.ars.usda.gov/.

Your hardiness zone is based on the minimal, historic cold temperature in the zone over a 30-year average. This is helpful information in determining whether a plant will survive extreme cold and how much winter protection it may need. It is not, however, as helpful determining what cultural practices to perform throughout the year, such as when to plant. You can be in different parts of the country with wildly different spring and summer weather patterns, but still be classified in the same hardiness zone because your coldest average yearly temperature is the same.

> Your hardiness zone is based on the minimal, historic cold temperature in the zone over a 30-year average.

Most experienced gardeners, myself included, know there is so much more to determining when is the right time to plant than a zone. With more extreme weather events—such as drought, freezes, and floods—happening around the world, many farmers and gardeners now use zone recommendations as a guideline only. Instead, they rely on good recordkeeping, intuition, and a little bit of trial-and-error to guide them to what is best for their farms and gardens.

Bare root roses are planted January through May depending on your growing zone. (See my chart on page 122, as a starting point for when to plant bare root roses in your zone.) Potted roses are planted April through October depending on your growing zone. In some locations like Zones 9 and 10, potted roses can be planted year-round. However, there are some exceptions to the best times to plant.

For example, flower farmers who plant in tunnels, greenhouses, or other sheltered environments that mimic conditions in warmer zones, will plant earlier than their usual zone-planting time. Every growing season is different, too. You may end up with a freak cold snap in the spring, or experience an early autumn heat blast, so regularly track your weather and local growing conditions to determine the best time to plant. You can also contact your local agricultural extension office, American Rose Society chapter, garden club, or join my online learning academy for additional resources to help you select the best time to plant roses in your location.

Soil Preparation

Get to know what you have to work with. The adage, "the best defense is a good offense," applies here, and you'll discover that preparing planting beds for your roses is no exception.

It's best for both gardeners and flower farmers alike to get a yearly soil test. (See the Resource Section for test kits I recommend.) Use the test results to guide how you improve soil with organic matter and other recommended nutrient amendments. The Resource Section also includes the recommended pH and soil nutrient requirements for garden roses. It's best to amend the entire planting bed or area prior to installing roses, rather than digging the planting hole and throwing a cocktail of stuff in the bottom of that hole. You may even find that your soil is perfect and requires no additional bed preparation at all. Remember, whether you are making additions to the soil or leaving the native soil as is, remove any weeds, old rose leaves, or other debris that can harbor insects and diseases from the area so you have a clean slate to work with before planting your rose.

While an annual soil test can guide your soil preparation decisions, it isn't absolutely necessary. If you don't want to go through the effort of annual testing, a good rule is to amend planting beds with a ratio of 1:3, high-quality, aged compost to 2:3 native soil in the top three-to-six inches. That recipe usually produces about two-to-four inches of amendments spread over the entire planting area. Just grab some organic bagged compost at the garden center or, for larger areas, order from a local bulk compost supplier.

It doesn't matter much whether you use animal, mushroom, or mint compost, but be sure whatever you're using is sterilized so it's free of weeds, well-aged to prevent burning, and you know the nutrient content of the compost material you are applying. Be conservative with your application, because without a soil test to guide you, less is more. You don't want to add more organic matter than you need because high levels of nutrients can have a negative effect on your rose's health and soil fertility. If you find you need more organic matter as the years pass, you can always add a top dressing of compost or mulch later.

> Remember, getting your rose in the ground is the most important step to take.

You also don't want to disturb the microbes, fungi, mycorrhizae, and other good beneficial flora and fauna living in the native soil that your new rose needs, so I'm not talking about renting a deep rototiller and ripping through one-to-two feet of soil. Imagine the compost like the icing on the cake of the native soil foundation that you need to spread on—it's best to lightly integrate the compost with a fork, shovel, or rake, so it mixes with the top three-to-six inches of native soil.

If you don't have time to prepare planting beds with compost, it's also okay to plant in native soil. I can't tell you the number of roses I've planted in my home garden with zero preparation, and they are growing healthy year after year. Remember, getting your rose in the ground is the most important step to take. If you find yourself putting off planting because you're short on time or can't get to the store to grab necessary amendments, just get the plant IN the ground. I promise that your rose won't die, and you aren't a bad plant parent!

The rose will adapt to its planting location, and you can top-dress the soil later with fertilizers, compost, or amendments as needed. I have seen far too many photos from customers who have a neglected cluster of bare root plants they purchased and left in the bag too long. The roses dry out or get moldy because the gardeners didn't have time to pick up compost or amendments.

Likewise, customers who purchase potted roses, only to leave them unwatered and forgotten in a dark garage, or with the pile of kid's toys on the side of the house, will soon find a dead, brown, rose plant. The best thing they could have done was to just get their roses planted quickly and worry about the other tasks later.

The tools you'll need to start a rose garden are commonly available, ranging from shovels and rakes to hoes of various sizes.

LET'S GET PLANTING

YOUR ROSE PERSONALITY

WEEKEND WARRIOR

Don't worry about fancy soil preparation or amendments if you don't have time. Select your rose and get it planted. You can worry about compost and fertilizers later. Do make time to water daily. Both bare root and potted roses need consistent watering during the first few weeks after you've planted them.

EVERYDAY GARDENER

Be sure to have mycorrhizae on hand before planting your rose. In all my years working with roses, this is the one extra thing I always make sure to do. I see beneficial results year over year with plants that receive mycorrhizae. They establish quicker and develop stronger, thriving root structures, and overall exceptional plant health.

ASPIRING ROSARIAN

Take the time to get a yearly soil test. It will help guide you to best prepare your soil to get your roses off to a great start. It will help your pocketbook too by keeping extra fertilizers and compost you don't need out of your garden so you can save that money for more rose plants instead!

PHOTOGRAPHY BY JILL CARMEL

Planting Tips

Dig the planting hole big enough to fit the root zone. Don't worry about a specific depth or width—the classic recommendation is two-feet deep by two-feet wide—but my best advice is to dig a planting hole big enough to fit the pot your rose came in, or to the depth of the bare root rose's roots.

Don't throw in everything but the kitchen sink. A quick search on the internet will give you so many recommendations—from bone and blood meal to fertilizers, and eggshells—that might make your head spin. If you add all of these "garnishes," it's like giving your roses a Long Island Iced Tea and the hangover to go with it. The plant won't know what to do with all of this, and will likely burn. The foliage will yellow, the growth could be stunted, and the plant may even die. Adding the kitchen sink can actually hurt the soil and the rose.

You can plant a rose even where an old one once grew. I've replanted hundreds of roses over the years in locations where prior roses existed. I've never once had an issue with the so-called "rose replant disease," which isn't an actual disease, but one of urban legend among rose gardeners. There's also no need to dig out and remove the old soil. I suggest pulling out your old rose, and amending the entire planting bed area with a thin layer of fresh compost or mulch. (You can also get a soil test as well for good measure.) Apply the mycorrhizae when planting and install your new rose. The one exception to the rose replant rule is if you know the soil has some type of contaminant that could have caused the previous rose to struggle or die (like glyphosate or other chemical residue). If that's the case, choose a different location until the chemical residue can be remediated.

To bury or not to bury the bud/graft union? That is the question! Again, a quick search of the internet and you'll find a grab bag of contradictory information. While there are a few outliers, most Rosarians are in the "bury the bud union" camp, regardless of the planting zone. Their rationale is that if the bud union is buried below the ground, it will be more stable in warmer areas, or during high winds, while plants with thinner, longer stems with top stock growing above ground will pull up and topple over because they are not anchored by the roots.

After years of experience, this is one planting rule where I go against the mainstream. I have found that when I bury my bud union, I get more suckers. Suckers are a nuisance, and are not desirable In all my years planting roses, I've

WHAT ABOUT SUCKERS?

Rose suckers are a nuisance, and are not desirable. A sucker is a rose cane or shoot that has grown from below the graft, or bud union, from the bottom stock of the rose. Suckers do not occur on own-root rose plants. So, how do you know if a cane is a sucker? This can be hard to tell on new growth, especially in the spring, but later in the year, it's easier as suckers always grow from below the graft.

Suckers look like long, wild/curvy, octopus-like tentacles arching from the base of the plant. They are usually taller than the rest of the regular canes, with leaves that are a different shape and color than the top stock. If your beautiful, large, white-blooming rose plant suddenly produces a burgundy flower cluster on a long, tall cane, the rose didn't suddenly change colors: you have a sucker on your plant that needs to be removed.

While cutting it off may seem like a good idea, it's only a quick fix. When you cut them off—just like with deadheading or pruning—it signals the plant to regrow again. The best way to treat suckers is to grab it with your hand at the base of the plant where they are connected to the bottom stock, and pull it down and off completely like you are ripping a Band-Aid in one quick motion. If you have buried your graft union, dig below the soil to find where the sucker originated on the rose below the graft and pull it off where the sucker cane connects to the base of the rose plant. Whether you cut your suckers or pull them off they still may come back, but with quick removal, the top stock of your rose will continue to grow and thrive, and not be overtaken by these pesky canes.

---

never lost a plant with the graft planted above ground due to wind or weather.

In warmer climates, I recommend leaving the bud union one-to-two inches above the ground for sucker prevention. If you do end up with suckers, above-ground placement also allows for easier sucker removal than when the graft is buried. The good news is there is no wrong way to plant your grafted rose. It will grow whether you're on Team Bury or Team Above Ground, so if you've always buried your graft and it works well for you keep on doing it.

If you didn't have time to get a soil test or fully prepare a planting bed with compost and amendments, that's okay. Be sure to remove any weeds prior to planting, and rake away old debris or leaves that could harbor diseases and insects. Dig your hole, plant your rose, and you're good to go. Just get your rose in the ground. That's the most important thing you can do.

WHERE, WHEN, & HOW

planting a bare root rose

The advantage of bare root roses is cost and convenience. Typically, they are affordable and by planting early, you get a jumpstart on the growing season. Here's how to plant a bare root rose.

INSPECTION

Upon receipt of your bare root rose, inspect the plant for any damage. Your rose likely will arrive with the entire rose placed moist in a plastic bag or with the roots only moist and wrapped in a sawdust burrito-like covering. If you are not able to plant immediately, wrap the rose in the plastic bag or sawdust cocoon that it came in and leave it in a dark, cool location, with a temperature of 35-42 degrees F (such as a garage or closet) for up to a week. If you are a flower farmer or floral designer with access to a floral cooler, you can store the sealed bag in the cooler. Do not leave it in a location where it will freeze. Check on it daily to make sure it remains moist, but that mold and fungus are not developing on the canes of roots. The longer the rose is held the more likely it will develop mold or decay of the canes and roots.

The number-one mistake people make is letting the rose dry out. Either they leave the bag open, or remove it from its sawdust cocoon, and the rose dries out and dies. You may not see the damage from drying out right away, but one-to-two weeks after planting, the canes will turn black and the rose eventually dies. In the bare root stage, your rose needs to remain moist and cool to ensure that it stays viable. Make friends with the mist setting on your garden hose nozzle and keep that rose moist and cool until you can plant it.

PREPARING THE ROOTS

Trim any broken canes or roots from the rose. Before planting, soak the roots in a large bucket or tub of water for 12-24 hours. This will help to rehydrate the roots and prepare it for planting. Choose a planting site that has six hours of direct sunlight everyday and rich, well-drained soil.

DIGGING THE HOLE

Dig a hole large enough to fit the root structure of the rose, which is usually about 12-18-inches deep and wide. Don't worry about a specific depth or width—the classic recommendation is two-feet deep by two-feet wide—but my best advice is to dig a planting hole to the depth of the bare root rose's roots.

IF YOU CAN'T PLANT WITHIN A WEEK

You can try a process called "heeling in" to hold the rose a little longer before planting. Heeling in is done by loosely burying most of the entire plant in a trench, hole, wheelbarrow, or bucket, and covering the roots loosely with moist soil. Check on the rose daily to make sure the soil and rose canes remain moist. While this can be an effective technique for holding roses before planting, it should be your last resort.

ZONE	PLANTING TIME
Zones 9-11	January-February
Zone 8	Mid-late February
Zone 7	Early-mid March
Zone 6	Mid-late March
Zone 4-5	Early-mid April
Zone 3	Mid-April - Early May

PLANTING

Build up a cone-shape mound of soil in the center of the hole. If you didn't have time to amend with compost prior to planting (as directed in my Soil Preparation section), you can throw in some compost and make your cone mound out of compost instead of native soil. This mound will protect the roots from breaking and promote natural downward growth. Sprinkle the roots of the bare root rose with mycorrhizae fungi. (See the Resource Section for sourcing mycorrhizae). Don't place any other amendments or fertilizer in the planting hole.

PLACING THE ROSE ROOTS

Place the roots over the cone like an upside-down martini glass. If you have a grafted rose, in colder regions, plant the bud union two-to-three inches below the ground. In warmer weather regions, plant the bud union one-to-two inches above the ground and lightly cover to the base of the graft with compost. For own-root roses, plant to the base of the canes, and then fill in and cover the roots with soil. Top dress the base of the plant with two-to-four inches of high-quality, aged compost or compost plus mulch. This will keep the feeder roots of your newly planted rose cool and moist during summer weather, stabilize the plant in the event of wind, and help control weeds during the first year.

POST-PLANTING CARE

Irrigate the plants daily for one-to-two weeks—longer if conditions warrant—with overhead water, a practice that keeps canes and roots hydrated. Depending on the weather and rainfall in your location, continue to water overhead on a daily basis until the first sprouts of buds start to push out from the canes. Once buds and leaves emerge, you can place the plants on a drip-irrigation system or hand-water at the plant's base as needed. For optimum rose health, do not apply any rose fertilizer until you see the first sets of leaves formed. (See Chapter 5 for fertilizer and ongoing watering recommendations.)

ALSO NICE TO KNOW

Climbing Roses

To put it simply, I love climbing roses! I love the way they surround an arbor like a welcoming hug, or trail along the facade of a building or fence line, reminiscent of a friend guiding me along a fragrant path. Climbers are versatile and ornamental. They add movement and texture to any garden space with blooms as diverse as their counterpart shrubs. Their foliage is not to be missed either, with deep, green-colored canes in the summer that fade to crimson, mustard, and burnt-orange shades as the seasons change.

Climbing roses can make great cut flowers too. With long, arching canes, they are wonderful for large-scale vase arrangements on an entry table, or for graceful wedding arch. Some of my favorite climbing roses that are ALSO great for cutting include: 'Polka', 'Sally Holmes', 'Eden', 'White Eden' and 'Butterscotch'.

Landscaping Roses

Since the days of ancient Greece, roses have been enjoyed by kings and queens, commoners, writers, philosophers, and gardeners alike. As humans evolved, roses went from a wild species to one of human cultivation. Used for everything from burial ceremonies to the seduction of a lover, in wild plantings on the side of mountains to formal English gardens, roses have been fixtures in the world's landscapes for millennia.

Today, modern rose varieties look beautiful planted in both formal rose gardens in collections all their own, or integrated into mixed landscape plantings where they can bloom among their flower friends. The roses that I specialize in growing at Menagerie Farm & Flower are roses that are both beautiful in the landscape, alone in a formal rose garden, or lined in rows at attention in a field for cut-flower production. They are the triple threat! These types of roses are usually classified as hybrid teas, floribundas, and grandifloras.

While any rose can be planted in a landscape, not all landscape roses are good cut flowers. What do I mean by landscape rose? While there isn't a formal definition of what a landscape rose is, they are classifications of rose cultivars that are generally

WHERE, WHEN, & HOW

planting a potted rose

Most people purchase potted roses with a few buds in bloom, which makes it easy to select plants that are true to the labeled variety. Here's how to plant a potted rose.

PREPARING

Before you remove the rose from its nursery pot, water it well 24 hours prior to planting. This will help rehydrate the roots, and keep the root structure intact when removed from the pot. If you water immediately before planting, the soil can become too moist and the established root structure can break apart. Choose a planting site that receives a minimum of six hours of direct sunlight every day and ideally has rich, well-drained soil conditions.

Dig a hole about three-to-four inches deeper than the pot, and six-to-eight inches wider (roughly three inches on each side). I don't recommend placing any synthetic fertilizer or other amendments, like bone or blood meal, in the planting hole.

To stimulate root growth at planting, sprinkle mycorrhizae fungi in the planting hole. Follow package directions on the mycorrhizae for exact application rate. If you're short on time this is not a necessary step, but I've found it useful in giving the rose a little jumpstart to healthy growth and establish roots.

PLANTING

Remove the rose from the pot and position it in the center of the planting hole. If you have a grafted rose, in cold weather regions plant the bud union two-to-three inches below the ground. In warmer weather regions, I recommend you plant the bud union one-to-two inches above the ground. For own root roses, plant to the base of the canes. If you are not sure if your potted rose is own root or grafted, position it so the soil is just at the base of the canes.

Backfill in the hole and cover the roots with soil, pressing firmly to remove air pockets (take care to not break roots). Cover the base of canes with compost or compost plus mulch to keep the feeder roots cool and moist in the warm summer weather.

POST-PLANTING CARE

Hand-water the rose at the base of the plant daily for one-to-two weeks to keep the roots well hydrated as the rose gets established in its new location. Newly planted roses need more water than established plants. After one-to-two weeks, place the plants on a drip-irrigation system, or hand-water at the base of the rose as needed depending on the weather.

No need to fertilize at planting as most potted roses you purchase from a reputable nursery will have already added a slow-release fertilizer to feed the plant for a few weeks after planting. You can begin a fertilizer regimen three-to-four weeks after planting when the rose has acclimated to its new location.

PHOTOGRAPHY BY JILL CARMEL

considered easy care, require minimal to no maintenance, will self-deadhead, and rebloom quickly. So, when someone asks me to recommend a good landscape rose, they usually mean something that is low-maintenance and isn't a cut-flower rose. For landscape rose recommendations, I look to roses that are classified as shrubs, ground cover roses, or the newer popular 'Drift' or 'Knock Out Roses'.

The great thing about landscape roses is that they really ARE easy to care for and require minimal to no deadheading to rebloom. They are the "set it and forget it" family of the rose world. If you are a Weekend Warrior just getting started with rose growing, these might be a great choice as you get your feet wet with minimal time commitment for care. You can design a beautiful cutting garden that includes landscape roses that will complement your cut-flower roses, perennials, annuals, and bulbs to create a design that will produce a year-round colorful display.

> The great thing about landscape roses is that they really ARE easy to care for and require minimal to no deadheading to rebloom. They are the "set it and forget it" family of the rose world.

Some of my favorite landscape roses are 'Tequila Supreme', 'Iceberg', 'Icecap', 'Limoncello', 'Peach Drift', 'Blushing Knock Out', 'Easy on the Eyes', 'Top Gun', 'Sweet Drift', 'Blanc Double de Coubert', and 'Bonica'. Tuck theses beauties in a hedge along a driveway or a border fence and they will be a welcome addition to the plant collection on any farm or garden. All my recommended landscape roses listed will do well in any growing Zone 4-10 (and even Zone 3 with proper winter care). Planting is easy, just like planting a cut-flower rose.

Guess what? These landscape roses can also be used as cut flowers too. While their vase life is fleeting with stems that are often are wild to tame, I cut blooms from time to time from the 'Iceberg' and 'Tequila Supreme' that are dotted around my yard to use in a simple bud vase on my desk, or to add a texture to a vase arrangement for a birthday celebration. So don't be afraid to snip the landscape roses, as in a pinch, they are cut flowers too.

Remember, any type of rose can be planted in your landscape so mix it up. Fill your garden with the textures and colors of hybrid teas, shrubs, climbers, and English roses, so your landscape isn't just a garden, but a colorful world of abundance for you to enjoy filled with roses of all shapes and sizes.

Planting in Containers and Raised Beds

It's so easy to grow roses in raised beds and containers. These options are ideal if you have poor soil quality, poorly drained soil, or aren't living in a location where your roses are going to be in their forever homes. All the roses in the Rose Gallery in Chapter 3 will do equally well planted in native soil as they will in a raised bed or pot. Be sure to check the care card on the rose before you select your bed or pot size, as climbing roses or varieties that have a wider span may require a wider pot.

Raised Beds

Raised Beds: Your raised bed can be as large or small as you'd like it to be. The sky is the limit on space. To build a raised bed you can use wood, brick, cinderblocks, stone, concrete, or really anything that can produce a firm sidewall without collapsing. Get creative with your design or you'll discover that a quick internet search will provide you with blueprints for constructing a raised planting bed. If you have gophers, moles, or other ground-dwelling pests, I recommend lining the bottom of the bed with chicken wire or metal mesh with a one-inch grid to prevent these pests from climbing into your bed.

Fill the bed with a mix of 50 percent high-quality, planting topsoil, and 50 percent compost. Bagged topsoil can be purchased at local garden centers and home improvement stores. Larger quantities, which you may need depending on bed sizes or quantities, can be ordered by the yard from a local rock yard or independent nursery center. Some nursery centers offer pre-blended mixes by the yard that already contain a compost-topsoil mix, so be sure to ask for an ingredient list when making your selection. Water the bed well after construction and filling. Getting your bed filled a day or two before you plan to plant your roses is ideal. If the soil settles after initial watering, you can add more soil mix before planting your roses. Be sure to start with a full bed of soil.

Containers and Pots

Containers and pots are also a wonderful way to plant cut garden roses. The process is almost the same as planting in a raised bed. From glazed ceramic to terracotta, and plastic to wood, you are sure to find a pot that is perfect for you.

transplanting a rose

You can transplant a rose during the dormant or growing season, although plants will adapt to a new location and experience less shock when moved during dormant season.

PREPARING

If you have embraced my "3R" planting mantra (Right rose, Right place, Right time), you likely planted your rose in the perfect location (wink-wink). But what happens if it isn't the perfect location?

Or, maybe you want more space between each plant; or you want to tighten them up? Or, perhaps you sold your house and want to take your roses with you. If you need to relocate your rose, it's easy!

Regular irrigation or watering is recommended to ensure success when you transplant a rose during any season of the year.

MOVING IN DORMANT SEASON

Aim to move any rose at the same time you perform dormant pruning. The steps for moving roses are almost exactly the same as pruning, and then planting a bare root rose. First, dormant prune your rose. Once completely pruned down to a 12-18-inch height with five-to-seven good canes, grab a spade and dig the rose out of the ground. Be mindful not to break the roots (although some root breakage is expected in order to get it out of the ground especially with an older, established rose that has a large root structure).

Once removed, shake off any excess soil until the roots are bare, and trim off any broken roots. You now have your own bare root rose that's ready to transplant in its new location. The steps are the same, whether your new planting location is in the ground or in a pot. Be sure to overhead water well the first few weeks (if you don't have rain), until the transplanted rose starts to push buds and leaf out. It's best to dig and plant on the same day.

MOVING IN GROWING SEASON

You *can* transplant roses during the growing season. Aim to transplant in the coolest weather during the early spring or fall, when the temperature is consistently below 80 degrees F. You can transplant in higher temperatures, but the rose will experience more stress and a higher likelihood of not surviving the move. To transplant in season, shape prune or trim the rose to a manageable height and width.
Once pruned, grab a shovel and dig the root structure out of the ground, leaving it as intact as you can, soil and all.

Your final location can be in the ground or in a five-gallon or larger pot. It's okay to trim away very long or broken roots so the rose will fit in its new location, just try to keep as much of the existing root structure as you can. Transplanting into a pot allows you to transport the rose if you are moving. Be sure to water well the first few weeks so the root structure can get reestablished in the soil.

PHOTOGRAPHY BY JILL CARMEL

Be sure to select a pot that has a drainage hole in the bottom. If it doesn't have a hole, depending on the material, you can drill three to four holes in with a 3/4-inch drill bit. Your pot also needs to be in a location where it can drain easily so if needed place risers underneath, using a block of wood. Or, add a saucer to keep your pot off the ground to allow for proper water drainage, and prevent standing water from pooling in the pot.

Make sure your container is at least two-feet wide by three-feet deep. Avoid pots with necks that are narrow—they make it difficult to remove the rose if you need to transplant it to a new container or location later. If you're located in a warm climate, you also want to avoid black or dark-colored pots, as they can absorb more heat, which can scorch roots in the summer months.

Fill the pot with a mix of 75 percent high-quality, planting topsoil and 25 percent compost. Bagged topsoil can be purchased at local garden centers and home improvement stores, and larger quantities can usually be found at a local rock yard or independent nursery center. Water the pot after filling. If the soil settles, add more soil mix before you begin planting your roses, so you start with a full pot.

Spring and fall are the best times to plant potted roses in beds and containers so they can get established before the summer heat or winter frost arrives. Bare root roses can be planted in pots during your dormant season. Remember to check on them regularly during the first few months to make sure they receive enough water. Roses planted in beds and containers dry out more quickly, and need more watering than their counterparts planted in native soil.

CAN YOU TIME YOUR BLOOM CYCLES?

Timing blooms for a specific week or occasion is best achieved after you have good recordkeeping notes, and a few seasons of collected data on the average time between flushes of roses for the varieties you have planted in your location. You can time the flush, or bloom cycle, by monitoring your roses and observing the period of time between deadheading or harvesting a rose plant for it to bloom again. Take that duration, usually about six-to-eight weeks, and count back that amount of time from the date of your event and deadhead or pinch back ALL of the buds on the plant.
It should then fully rebloom just in time for your special event.
This isn't a perfect science as weather, water, plant variety, and other factors can hasten or slow the process, but with good recordkeeping and bud-pinching, you can try your hand at getting your roses to bloom just when you want them to.

5

IN THIS CHAPTER

CARE PROGRAM OVERVIEW

DETERMING THE TYPE OF CARE

MULCHING & COMPOSTING

FERTILIZING

PEST & DISEASE CONTROL

WEEDS

WATERING & IRRIGATION

PRUNING

CARE CHART

YEAR-ROUND ROSE CARE

When I started producing my field-grown garden roses as a commercial farmer, I soon realized I was on my own. There weren't any online resources or reference books that could guide me on growing these beloved plants as cut flowers. When I searched library resources, I found informational texts on growing greenhouse varieties and hydroponic rose farming. Millions of Google hits about growing garden roses for show gardens or rose exhibitions appeared, but the practical help I sought was nonexistent.

As I started my journey as a professional grower, I tried many cultural practices and methods—from the hydroponic world to the rose show exhibitor methods of the 1980s, 1990s, and early 2000s. I read all of the rose lore, including advice like water deeply once a week; don't cut stems off a first-year bare root plant; harvest when the bud has popped; pinch all new buds each season, and more. I was left disappointed as none of the methods for variety selection, pest control, irrigation, or amendment applications provided me with the quality of field-grown, cut garden roses that I wanted to grow.

So, after my first very frustrating year, I returned to the drawing board and used my professional farming expertise as a crop scientist. I wanted answers to these questions: What happens if you space the plants tighter? What if you water more frequently? How can you manage insects and diseases with minimal intervention effectively? Can you cut deeper on a first-year plant? I went back to basics, from my college courses in soil science, plant pathology, and entomology, with a little family knowledge thrown in, too.

Much of my success with growing garden roses is based on a regular habit of plant care and maintenance. The payoff for my effort and attention comes during bloom season.

After all, I had my mother's and grandmother's traditional ways, as lessons from their rose gardens also shaped my approach to growing garden roses as cut flowers.

I also remembered after that first challenging year, that farming and growing should be enjoyable. It will never ever be perfect, but I've learned to live in harmony with the inevitable presence of insects and diseases, and to adapt to weather anomalies that occur. I accept the flaws, but I don't think of them as imperfections. Just as my grandmother taught me all those years ago, I truly see the distinction of each rose, and know that its mark, shape, size, and color contribute to its unique character. And that should be celebrated. In the same way that viticulturists describe a grape's distinct environment as "terroir," which includes the soil, topography and climate, I believe each rose has its own terroir as well, influenced by the unique environment in which it's produced.

I'm holding two favorite roses in the golden and butter palette: 'Graham Thomas' and 'Charlotte'.

Every rose has characteristics of color, shape, size, and fragrance imparted to it by the environment in which it is produced. The roses you grow are as unique as you are. In this book, I've gathered all the garden rose care information that I wish I found when I started growing. These practices have set up both my roses and Menagerie Farm & Flower for success, and I know they will help you discover the special terroir of your own rose garden.

CARE PROGRAM OVERVIEW

How and when we care for our roses is a bit like customizing a personalized exercise and nutrition routine. I know my doctor says to eat healthy, but I enjoy my desserts and glasses of wine. Likewise, I also don't like how I feel if the only exercise I can manage is a half-hour each week dancing around the house with my kids, and eating fast-food dinners on the way home from school pick-up. I'm happy somewhere at the mid-point of the self-care spectrum that balances health and the joy of life's everyday pleasures.

Like a workout or meal plan, choose a rose-care program that will work for you and your plants. Throw away any unrealistic expectations you may have read on Dr. Google, or seen on the numerous social media forums about roses and flower farming. Or even in this book! Much like exercise and nutrition, when it comes to roses, the rewards are commensurate with your efforts, and you will be successful when you find a program that works for you.

Even with the best-laid plans, I often fall off the healthy-life wagon, and after years of feeling defeated, I now shrug it off and remember that tomorrow is another day. So, choose a care program that fits your lifestyle and don't feel guilty if it all doesn't get done. No time to compost? Don't. Can't monitor for insects each week? Don't worry and let them feast. Forget to fertilize? It's okay. Do what you can with the time you have.

If you aren't getting the results you want, then consider reallocating your time or taking on less. Yes, less. Be brutally realistic. Don't give up on roses completely, just scale back to a collection and maintenance tasks that are sustainable for you. I've had many students in Menagerie Academy during the past few years who want help planning a new rose farm or cutting garden. Some even have wild dreams to plant anywhere from 50 to 3,000 roses, having never farmed before.

However, after we evaluate their weekly allocation of time and resources, they usually decide to plant more roses than the time or experience allows for successful results. Many still plant more roses than advised and often struggle.

The lesson here is: Don't bite off more than you can chew! You can always phase in more care steps and more rose plants as you develop the program that works for you. Remember, rose growing should be enjoyable as you fit the beauty of roses into your life.

DETERMINING THE TYPE OF CARE

The goal of having a home cutting garden, or adding roses to your flower farm's collection of plants, is not just to have a beautiful display of blooms, but to have a consistent supply of rose stems to cut. This is a different goal than building a formal display garden, rose-filled residential landscape, or exhibition roses. This small but important distinction was why the advice from older rose-growing books didn't help my growing goals. There's an old saying in the flower-farming world: If you see a field filled with colorful flowers in bloom, it means the farmer isn't making any money, as almost all flowers should be cut before their buds fully open.

I find most people looking to have a rose-cutting garden are somewhere in the middle. They want to have an ongoing supply of stems to cut, but they also enjoy having an abundance of rose blooms on display for all to enjoy. That happy hybrid is entirely achievable!

The care methods I've honed over the years are a balance of efficiency in cultural practices, scientific research, and testing fact vs. fiction of old wives tales. To grow a successful rose-cutting garden, I encourage you to take the care information in this book and use it as template. Start to develop your own care practices and you'll find, as I've found, the ideal approach that works best for you and your cutting garden goals.

Monitoring Your Roses for Care

Whatever you call it—journaling, note-taking, or recordkeeping—it's one of the most important things you can do to grow garden roses successfully. Recording the when and what you do in the garden requires minimal effort, and is essential to track patterns that emerge as seasons change, giving you a historic reference to compare a rose variety's performance year over year. Gathering data guides your decisions about your practices, as well as helps document which roses are superb performers and which ones need to be retired to make room for a happier variety.

To begin, develop a consistent plan to document what is happening on your garden or farm. What time of day will you observe? How frequently can you check plants—daily, weekly or monthly? What do you want to document? Will you use an old-fashioned paper and pencil? (By the way, that's my method of choice for note-taking when I walk the fields, but I transfer those notes to a computer

YEAR-ROUND ROSE CARE

file later.) Or, use the notes app on your phone, build a spreadsheet or any other combination of data gathering that works for you.

In addition to written documentation, don't forget to use your camera to take photographs and videos. The phone in your pocket makes it easier than ever to snap a portrait of a beautiful bloom or photograph unwanted pest; then store the images by month or category in your photo file archives.

THINGS TO DOCUMENT

- Weather: high and low temperatures
- Rainfall
- Soil test results
- Insects observed
- Diseases observed
- General appearance of the plant (great, okay, or needs attention)
- Irrigation application rates and dates
- Fertilizer application rates and dates
- Pest and disease control applications rates and dates
- When you deadhead
- Time from deadheading or harvesting to next bloom
- When you harvest stems
- How many stems you harvest
- When you dormant prune
- When you winterize (in select zones)

If time is a factor for you, don't document everything. Choose the top three things you want to document, or split up the items you are collecting during different times of the week or month. Farm life and home life can get very busy for me, so I use task management software on my computer/phone calendar to ping me with reminders to do my observation and recording.

When I look back at the years of documentation I've collected, I'm so glad that I keep my journaling appointments with my roses. Whatever frequency you choose, stay consistent and over time, you will see patterns develop. You'll discover when the most amazing blooms of the season appear, when the insect pressure gets bad, or when to expect yellowing on the leaves in late summer. Your journaling of positives and negatives will be unique to your location, and provide you with a roadmap for the future as you build a thriving and successful rose-cutting garden.

MULCHING & COMPOSTING

There are many benefits to proper mulching and composting. As compost decomposes, it feeds the soil, increases biodiversity, and improves soil structure. Layers of mulch keep roots cool during warmer months, enhance water infiltration and retention, and suppress weeds by shading their seeds from germination; The list of benefits is numerous.

So what's the difference between a compost or mulch? You may see both of these terms often used interchangeably, but there is a subtle difference between them. Compost is a material that is already broken down from its original organic composition or decomposed. Grass clippings, leaves, yard waste, straw, bark, and wood chips are all types of mulch. Compost is what becomes of material, like those listed above, after it breaks down and can include food waste, animal waste or even yard waste.

> Compost has a dual beneficial role as it emits nitrogen into the soil and into the air.

Mulch is a material that you dress over the top of the soil. It has been collected from an original source and hasn't broken down from its original composition yet. Mulch needs something to assist in the decomposition process. That special something is nitrogen. Mulch draws the nitrogen it needs from the soil to break down, so if you apply mulch (wood chips, straw, grass clippings, leaves) that layer will remove the nitrogen from the soil to start decomposing. This is the opposite of what most growers want!

We want to feed nitrogen to our roses from the soil rather than remove it. Yet once mulch has broken down and turned into compost, then the opposite process happens. The compost feeds the soil with nitrogen. Compost has a dual beneficial role as it emits nitrogen into the soil and into the air. So why do people throw wood chips around the base of their plants if it's taking nitrogen from the soil? That's a good question and my answer is that most well-meaning gardeners probably don't know that until the mulch they have applied decomposes, they are actually stealing nitrogen from the soil. This is also why many gardeners who aggressively apply wood chips as mulch find themselves needing to apply fertilizers high in nitrogen to compensate.

WEEKEND WARRIOR

BASIC CARE PLAN

Journal and document the performance of roses on a monthly schedule

Dormant prune

Apply dormant oil and fungicide spray

Apply compost/mulch once after your dormant pruning or early in the spring

Apply granular fertilizer when first leaves appear

Reapply granular fertilizer mid-season

Monitor for pests as time allows

Apply insect spray as needed

Check irrigation and soil moisture on a monthly schedule

Deadhead as time allows

Harvest stems as time allows

Winterize (if needed)

ENJOY YOUR ROSES!

EVERYDAY GARDENER

BASIC CARE PLAN

Journal and document the performance of roses on a weekly schedule

Dormant prune

Apply dormant oil and fungicide spray

Apply compost/mulch once after dormant pruning and mid-season

Apply a granular fertilizer when first leaves appear

Reapply granular fertilizer every six-to-eight weeks

Monitor for pests and diseases on a weekly schedule

Apply insect or disease spray as needed, every two-to-three weeks

Perform weed control weekly

Check irrigation and soil moisture weekly

Deadhead at least once weekly

Shape-prune in mid-season

Harvest stems twice per week

Winterize (if needed)

ENJOY YOUR ROSES!

ASPIRING ROSARIAN

BASIC CARE PLAN

Journal and document the performance of your roses daily

Dormant prune

Apply dormant oil and fungicide spray

Apply compost/mulch three times per year after dormant pruning, mid-season and fall

Apply a granular fertilizer when first leaves appear

Reapply granular fertilizer every six-to-eight weeks

Monitor for pests and diseases daily

Apply insect or disease spray as needed every one-to-two weeks

Apply foliar fertilizer every two weeks

Perform weed control twice per week

Check irrigation and soil moisture twice weekly

Deadhead daily

Shape prune in mid-season and fall

Harvest stems daily

Winterize (if needed)

ENJOY YOUR ROSES!

So what should you choose: Mulch, compost, or both? After years of trying a number of combinations, I've narrowed down my mulching and composting to two different programs. I usually spread compost two-to-three times per year in the late winter, mid-season and early fall. In your home or garden, the number of applications you perform may only be once or twice, depending on the length of your growing season. For a basic care plan, I recommend that any rose grower plans at least one application of compost or compost plus mulch annually, best scheduled just after dormant pruning. (In colder regions, be sure to apply after your ground has thawed so soil warming isn't delayed.) Remember, before applying any organic matter, be sure to have a soil test performed. If your soil is already high in nitrogen and organic matter, you may not need to apply anything. Lucky you!

There are so many different types of compost and mulch that a full review in this book would be impossible. So I recommend finding a local nursery or agricultural supplier to provide you with the options that are successful in your region. Whatever you choose, mulching and composting roses is essential for a balanced rose care plan. Here are two approaches I recommend:

FOR A LANDSCAPE CUTTING GARDEN: I take a two-step approach, as I want the benefits of both the compost and mulch. I first apply a two-to-three-inch layer of compost to transfer essential nitrogen from the compost into the native soil and then top the compost with a two-to-three-inch layer of mulch, choosing material like wood chips, leaves, or straw. This dual action prevents weeds, feeds the soil, and provides a decorative finish that looks attractive. It also provides a base layer for the soil so the mulch, which needs nitrogen to break down, is drawing it from the fresh compost and not from the soil. It's like a perfect nitrogen-emitting and nitrogen-decomposing sandwich.

FOR A COMMERCIAL ROSE FIELD: In my rose field, I take a single-step approach, and only apply compost in a two-to-three-inch layer, two-to-three times annually. In my fields, we are very aggressive with daily care, hand-cultivation because I have high weed pressure. We are constantly weeding with hoes and performing other maintenance tasks, so any mulch we might apply usually gets removed from the rose beds or spills into the walkways before it has time to break down. The thicker layers of straw or wood chips actually make it more difficult to perform maintenance tasks, and leaves a messy-looking field. So, if you want to save a few pennies, just apply compost as needed, and save the mulch for the landscape. Or, if you have lower weed pressure—or like the look of mulch in beds—you can stick with method one.

FERTILIZING YOUR ROSES

Applying fertilizer to your roses isn't necessary. I know! That's a shocking statement, as many of you may have heard that roses are "heavy feeders." The truth is, if you have a good soil profile and compost yearly, then your roses will survive without fertilizer, or at most, require just a single application in the spring.

In a rose-cutting garden or flower farm, though, we aren't growing regular landscape roses. Our goal is to grow beautiful roses with long, strong stems for cutting. So the reality is, in a cutting garden or flower farm you will need to apply fertilizer to help the roses perform their best and bloom prolifically.

So what are the best types of fertilizers to apply? There really is no best fertilizer, but there are fertilizers that perform differently based on the percentage of ingredients and method of application.

Determine your rose-feeding program over the course of the year. Granular fertilizer is quick and easy to apply.

Types of Fertilizer

There are two types of fertilizers: Organic and synthetic. Organic fertilizers are derived from living organisms like bone meal, alfalfa meal, fish emulsion, and blood meal. They typically have a low percentage of nitrogen and release their nutrients slowly, preventing any burn to the plant. Synthetic fertilizers are chemically manufactured and are not derived from living organisms. They are generally higher in nitrogen and are often fast-acting, so they can burn plants if applied incorrectly. When applied correctly they are effective.

Methods of Application

The goal of any fertilizer program is to help maintain not just a healthy plant but healthy soil. As I discussed in Chapter 4, getting a yearly soil test is an essential place to start. A soil report will tell you if you even need to start feeding. Use those results as a jumping-off point to determine the best type of application program for your roses, as you may not need to apply as frequently as my method.

Fertilizers can be applied in liquid, foliar, or granular form. Liquid fertilizers are mixed with water and applied to individual plants at the base of the plant. If you have a drip line installed with the necessary valves, and a backstop and injection system, you can add liquid fertilizer through the water lines in a process called fertigation. Foliar fertilizers are applied by mixing the nutrient solution with water and spraying the leaves and canopy of the plant. Dry fertilizers, or granular fertilizers, are applied at the base of the plant by sprinkling small granules around the base of the plant in the root zone.

All three types of fertilizers can be found in local nursery centers as all-in-one products that contain not just fertilizer, but insecticide and fungicide too. These all-in-one products can be beneficial for the grower who is short on time and

HOW TO READ A FERTILIZER LABEL

The label on any fertilizer will have three numbers listed which correspond to the percentages of nitrogen (N), phosphorus (P), and potassium (K), or N-P-K. The numbers and elements always appears in the same order on the label. For example, a fertilizer of 10-8-8 equates to 10 percent nitrogen, eight percent phosphorus, and eight percent potassium, by weight. A quick read of a product label will also show you the ingredients from which the fertilizer was derived, such as alfalfa meal, kelp, and other materials. It also lists any other ingredients included in the blend, such as mycorrhizzae, that can be beneficial to your soil and roses. High-quality fertilizers from reputable brands are well formulated, so anything you choose from your local garden center or online retailer will help your garden roses shine!

wants a one-stop solution. With any fertilizer, be sure to read the ingredients and instructions on the label to ensure you are applying the appropriate rate of product, and be sure to wear appropriate Personal Protective Equipment (PPE) as required.

LIQUID: Use this method if you have a just few (under 20) rose plants, as it can be tedious to mix and apply liquid fertilizer to each plant. Save the liquid applications for farms or gardens that have drip or injection systems in place.

FOLIAR: Spray until the foliage is glistening, but not quite dripping. Be sure to get the undersides of the leaves as well. Don't apply when the weather is hot, because temperatures above 80 degrees F can cause the fertilizer to burn the leaves. If you have fewer than 50 roses, use a basic one-to-two-gallon, hand-carried spray tank. If you have more than 50 roses, consider investing in a gas- or battery-powered backpack or tank sprayer.

DRY: Apply to dry soil around the base of the rose plant, covering the area equivalent to its root zone (the radius around your plant comparable to the spread of its branches). You may need to scrape away the mulch layer in order to apply fertilizer; then redress the top with mulch. Granular fertilizers are best applied just before a rain, as they need water to be activated. If there is no rain, you will need to thoroughly hand-water each plant where the fertilizer was applied.

Fertilizing the Menagerie Way

This is a simple, all-in-one program you can schedule into your calendar. It works well for both a home cutting garden and rose flower farms. If you are a Weekend Warrior who doesn't have much time, just omit an application. If you are an Aspiring Rosarian and want more productivity, schedule foliar applications closer together, or incorporate a weekly fertigation program. I recommend applying a granular slow-release fertilizer every six-to-eight weeks throughout your growing season. (You could also can apply liquid in place of the granular.) Begin your first application when the rose has just started to leaf out. Plan to apply your last application six weeks before your first expected frost. In the spring, choose a fertilizer with more nitrogen to encourage new growth. By mid-summer, choose fertilizer with less nitrogen and more phosphorus.

Also, make sure to water the granular material into the soil if you are not expecting rain, or be sure the application aligns with the drip line where it will receive water. A great way to remember when to apply a granular fertilizer is to use the major U.S. holidays: Spring Break (Easter/Passover), Memorial Day, Fourth of July, and

Labor Day. Be sure to incorporate a foliar spray into your program every two-to-three weeks.

All fertilizers, whether naturally-derived or synthetic, are comprised of the same three elements: nitrogen, phosphorus, and potassium. Become a careful label-reader when you visit the garden center, and choose wisely for your rose care program. (For a list of my favorite fertilizers see the Resource Section.)

PEST CONTROL

Insects are my number-one nemesis here at the farm, causing me the most sleepless nights. Questions about handling insects are also by far what regularly fill up our email inbox. People want to know: What's eating their roses?! We all feel that gut punch when we head out to harvest roses only to find them dotted with thrips, or see a colony of Japanese beetles has taken up residence.

That said, thankfully there are good guys, too! There are many beneficial insects that are predators of harmful ones, so it's important to sustain an environment where they can thrive among your roses. Maintain an active scouting and monitoring program, encourge beneficial insects, and selectively use insecticides.

There are so many insect species, ways to monitor, and methods of control that it would be impossible to cover them comprehensively this chapter. Honestly, I could write an entire book on insect and disease control in roses. My favorite mentor and college entomology professor, Dr. Jo Ann C. Wheatley, at Cal Poly San Luis Obispo, always said, "look for the good guys first." In my Menagerie Learning Academy, I provide customized insect-control programs for members, and teach them to identify insects and look for the "good guys"—the beneficial insects that help provide a sustainable approach to integrated pest management.

Unfortunately, I've found that after years of treating insects on my own farm and for my clients, there is no one-size-fits-all when it comes to pests. I know you would like a magic recipe from me for treatment, but the most troublesome pest to you will depend on your growing region. So for the purpose of this section, I am going to cover the top methods of monitoring, identification, and control for the most common insect issues that my customers and Academy members encounter.

STEP ONE: MONITOR. The first step in a good insect-control program is monitoring. You don't want to apply a spray if you don't need to, but you do want to be ready to react at the first signs of a problematic bug. To monitor bugs, you

It's easier to visually identify pests on the undersides of leaves with a hand-held magnifying tool.

Pest Control Basics

can use several different methods from hanging sticky traps to catch pests, to daily rose inspections with a hand lens.

In short, you have to know what you're up against before you can treat the problem. Inspect your roses weekly or as often as you can. Look at the interior of the buds, pull apart the petals, remove leaves, and use a hand lens to inspect them from top to bottom. You'd be surprised at what you can find with a quick visual inspection under magnification.

With frequent inspections, you'll be more likely to catch an infestation early, before there's significant damage caused. Not sure what's damaging your rose? Take a sample of the plant or send a photo to your local agricultural extension office or master gardener organization in your area. This identification service is usually free, and a great way to connect with fellow gardeners and crop advisors near you. I also offer this service in 1:1 consultation sessions through my Menagerie Academy.

STEP TWO: TREATMENT. You found a pest! Now, what do you do? Choose a treatment option. You may choose to leave things alone and wait for nature to correct itself. You can also elect for a non-invasive treatment (water hose or hand-picking); introduce natural predators; or you can apply a pesticide (organic or non-organic). (See my list of common rose pests and treatment options below.)

Always do the least-restrictive, non-invasive treatment first and gradually get more aggressive from there, if required. For example: Wash off aphids with a spray of water before you reach for a chemical control method.

Inspecting a rose bloom that shows signs of pest infestation. Thrips are insects that resemble tiny wood-splinters.

STEP THREE: APPLY TREATMENT. Execute the treatment and go back to step one to continue monitoring. While there are many methods of controlling insects from cultural practices, as well as chemical (organic and non-organic) controls, I am sharing a few of my favorites that I've found to be most effective.

Rotate the insecticides (organic or non-organic) you use so your insect population doesn't develop resistance to any one formulation. A good rule of thumb is to use a product for two applications in a row; then rotate to another product with a different active ingredient. Rotate between three-to-four products and modes of action before you start the rotation again.

Spray insecticides, organic or non-organic, during the coolest part of the day, preferably in the morning before temperatures reach above 80 degrees F when pollinators are less active.

If you are a flower farmer looking to apply insecticides, even if the products are organic and available from a local nursery, be sure to obtain the proper county or municipal permits for commercial use.

Take thorough notes about pests and the control methods you use throughout the year, and your pest-control care program will improve.

Common Pests and Treatments

THRIPS: This insect is almost microscopic and, depending on the species, may not be visible to the naked eye. (These little devils are my number-one nemesis!) Upon visual inspection, thrips often look like small wood splinters sprinkled on the petals. They like to hide deep in the petals of a tight rose bud, deforming it by rasping and puncturing the tissue. Thrips are most active in the early spring, attacking the tender growth of emerging rose buds while the first flush is blooming. Treatments include removing infected buds; introducing the predatory mite Amblyseius swirskii; and spraying with neem oil, spinosad, or other chemical insecticide.

APHIDS: They come in all colors and sizes: green, black, red, white, and brown. They are a soft-bodied insect that likes to feed on new growth, so they are most active in the spring, or after you deadhead and new growth is sprouting. They love new red growth. Treatments include hosing off with water; squishing them with your fingers; encouraging a ladybug population as a natural predator; and spraying with neem oil, spinosad, or other chemical insecticide.

SPIDER MITES: Tiny pests that look like spiders, complete with webs, only smaller. The suck the juices from the leaf surface leaving dry, shriveled, and see-through leaves. A quick check to the underside of the leaf and you can see the telltale webbing. Spider mites most commonly appear in dry, hot climates, as they hate moist environments. Treatments include hosing off with water (they hate water!); spraying with neem oil; or other chemical insecticide.

ROSE LARVAE, CATERPILLARS AND WORMS: Small larva of flying insects like moths that chew holes in the leaves of roses. They are usually found on the underside of leaves and are more active at night. Don't be afraid to go out with a flashlight in your PJs to search for these critters. Treatments include hand-picking, and spraying with neem oil, spinosaid, or other chemical insecticide.

JAPANESE BEETLES: Shiny bugs that emerge from larva out of the ground during the midsummer. They can devastate a rose garden by feeding on the buds and leaves. Treatment methods aren't one and done, as tackling Japanese beetles requires consistency and patience over a period of years to gain control, but you will never be rid of them completely. Treatments include hand-picking, applying milky spore, and spraying with neem oil (which is less effective when large numbers are present), spinosad, chlorantraniliprole (low-risk residual with minimal risk to pollinators), or other chemical insecticide.

INSECT CONTROL GLOSSARY

CONTACT INSECTICIDE: An insecticide (both organic or non-organic) that will kill an insect or its larva/eggs only on contact. It provides no residual or long-lasting protection within the plant.

SYSTEMIC INSECTICIDE: An insecticide that will enter the tissue of the plant and kill the bug when insect eats the plant. Often long-lasting, it provides two-to-six weeks of residual insect protection.

Organic Insect Control Methods

HORTICULTURAL OIL: Refined from plants or petroleum depending on the product. This is a contact insecticide that smothers an insect and its eggs.

NEEM OIL: A botanical insecticide that is derived from the bark of a tree. Neem is a contact insecticide that repels pests and causes them to stop feeding.

SPINOSAD: Made from a soil bacterium, it is a contact insecticide that is a mixture of two chemicals called spinosyn A and spinosyn D. It affects the nervous system of insects that eat or touch it, and causes their muscles to flex uncontrollably, which leads to paralysis and death, typically within two days.

PREDATORY INSECTS: Insects that feed on other insects; They are also often called beneficial insects. If you choose to introduce predatory insects, you may not be able to spray insecticides—either organic or non-organic—as any insecticide may also kill your beneficial predators. Check with the insectary supplying beneficial predator insects for more information before any chemical application.

Non-Organic Insect Control Methods

ALL-IN-ONE: Sold under different brand names, such as BioAdvanced All-in-One or Bonide All-in-One, these rose products contain different active ingredients in combination (fertilizer, insecticide, and fungicide.)

Available as granular, foliar, and liquid. Read the product label to see the active chemical ingredients in each brand. Usually systemic products depending on application type.

TALSTAR P (BIFENTHRIN): A man-made version of a pyrethrin, which comes from chrysanthemum flowers. As a systemic product, Bifenthrin interferes with the nervous system of insects when they eat or touch it.

CARBARYL: A chemical man-made pesticide that is toxic to many pests. When insects eat or touch carbaryl, it over-stimulates their nervous system, causing a chain reaction including a disruption in their production of enzymes which causes death.

Vertebrate Pests

Almost as bothersome as insects, vertebrate pests can wreak havoc on a rose garden. The most common ones causing heartache to rose growers are gophers, moles, deer, and rabbits. Unfortunately, I don't have a magic bullet for controlling any of these pests, so I'll give you the high points for each.

GOPHERS, MOLES AND VOLES: They can take down the root structure of a rose in a flash. Planting roses in wire mesh baskets is about 75 percent effective, as the burrowing creatures will often burrow right up beside the basket and crawl up and over into the base of the rose. For a really bad infestation, invest in a gopher gassing machine.

DEER: Enclose your roses with a high fence. Whatever height you are thinking of building, double it! Deer often find a way over even tall fences, and they can eat a rose to the ground in a single feeding. The fence can be electrified as an added measure. Chemical control repellants are also available to spray around the perimeter of roses.

RABBITS: Much like repelling deer, you may need to invest in a fence to keep them out. You can also set traps to catch and relocate. Chemical repellants are also an option, but often not very effective.

DISEASE CONTROL

My philosophy of disease control is much like my approach to insect control. The reality is that diseases are always present and cause stress to even the most calm rose grower, and much like insects in the rose garden, you will never be able to rid your roses of all diseases. The number-one way to fight disease is to have a healthy plant, so keeping roses well-irrigated and maintaining healthy soil will improve your plant's ability to fight disease. Just like a human fights a cold with a healthy immune system, a healthy rose can withstand infestations of diseases.

> The number-one way to fight disease is to have a healthy plant, so keeping roses well-irrigated and maintaining healthy soil will improve your plant's ability to fight disease.

The most troublesome rose diseases you may encounter depend on your growing region. However, the BIG FOUR common ones in North America include: Black spot, rust, powdery mildew, and downy mildew. Unlike insects, these diseases are all basically treated the same way with the same materials. The less common—botrytis, canker, and anthracnose—can also be treated similarly. Fungal diseases are simply spores floating around in the air, and when they come in contact with the rose leaf, a moist environment, and the right weather conditions, the spores take residence on your rose and multiply. Why should you treat fungal disease? They are not only an eyesore in the rose garden, fungal infections can also cause enough leaf damage to defoliate the plant, leaving no surface area for it to intake light to perform photosynthesis and survive.

With any disease-control program, be sure to read all labels to ensure you are properly applying the product at the correct interval between applications, and always wear the proper PPE. The number-one reason most people fail at disease control is that they don't apply a product properly and according to directions.

Research whether your local agricultural extension or master garden program offers a chemical application safety class. Learn how to read labels, properly protect yourself, and mix and apply chemicals (organic and non-organic), and build the skills that will enhance your rose growing, gardening, and farming knowledge. Remember, even if a product is labeled for use in organic production, it's still a chemical and needs to be applied properly and safely every time you use it.

YEAR-ROUND ROSE CARE

Inspect rose foliage for telltale signs of disease.

Disease Control Basics

STEP ONE: PREVENTION. Avoid overhead watering, create a dry environment with ample airflow between plants, pick off infected leaves from the plant and discard in trash (do not compost). Remove any infected leaves or debris from the base of the plant, and apply fungicide as needed. Plant disease-resistant rose varieties that have natural immunity built into their genetics.

Practice good hygiene and clean your tools regularly before pruning or cutting; otherwise, if you have a disease infestation, you may spread it from one plant to the next. Don't skip your winter spray schedule and clean the rose bed of debris after dormant pruning. The single best way to prevent disease is to apply a combination of horticultural oil and fungicide in the winter, giving your roses a healthy start to the growing season.

STEP TWO: MONITOR. You don't want to employ a disease-control method if you don't need to. Take good notes about diseases and the control methods you used throughout the year, and your control care program will improve every year.

STEP THREE: TREATMENT. Always do the least restrictive, non-invasive treatment first and get increasingly more aggressive from there. For example: Spray with an organic treatment before you reach for a non-organic chemical control method.

Remember to rotate the fungicides (organic or non-organic) you use so the disease population doesn't develop resistance to any one chemical. A good rule is to use a product for two applications in a row, then switch to another product with a different active ingredient. Rotate between three to four products and modes of action before you start the rotation again.

Spray fungicides, organic and non-organic, during the coolest part of the day, preferably in the morning before temperatures reach above 80 degrees F, and pollinators are not active, as some disease sprays can harm pollinators.

If you are a flower farmer looking to apply fungicides, even if the products are organic and available from a local nursery or garden center, be sure to obtain the proper county or municipal permits for commercial use.

Common Diseases and Treatments

BLACK SPOT: As the name implies, this disease displays as circular black spots often with a ring of yellow around the exterior. It thrives in damp, moist weather, infecting the leaves first and then moving to the canes. Arguably the most common disease in roses.

RUST: Orange fuzzy spots on the underside of the leaves that look, as the name suggests, like rust or a rusted metal surface.

DOWNY MILDEW: Irregular purple to reddish spots advancing to yellowing of the leaves. In advanced infections, the leaves will turn brown and defoliate. Downy mildew can spread quickly in just a few days so start a control method immediately.

POWDERY MILDEW: White fuzzy patches that coat the leaves, stems and flower buds. Common even in dry regions in the early spring after a rain, or period of cloudy, humid weather. Even in my arid climate, I have mildew infestations on overcast spring days and in areas with poor air circulation. In advanced infections, the leaves can look crumpled like a piece of discarded paper.

Organic Fungicide Control Methods

HORTICULTURAL OIL: Refined from plants or petroleum depending on the product. This is a curative fungicide and will smother disease spores.

NEEM OIL: A botanical insecticide that is derived from the bark of a tree. Neem

> **DISEASE CONTROL GLOSSARY**
>
> **CURATIVE:** A fungicide (both organic and non-organic) that will attack fungi already infecting the plant. It provides no residual or longer lasting protection within the plant. For use on roses after an infection has taken hold.
>
> **PREVENTATIVE:** A fungicide (both organic and non-organic) that will inoculate plant tissue so that the plant is better able to fight off disease. For use on roses before an infection becomes bad.

is a preventative fungicide and repels the disease spores.

SULFUR: Can be applied as a dust or mixed as a liquid, and kills on contact.

COPPER: Great to use in season and as a dormant spray. Available as a liquid product, copper is my fungicide of choice here at the farm. Be careful with application as copper can burn; always wear eye protection and proper PPE.

CEASE: Is a contact biological fungicide containing a patented strain of the bacterium *Bacillus subtilis*. First, the bacterial spores occupy space on the plant surface and compete with the pathogens. Next, active compounds lipopeptides produced by each bacterium disrupt the germination and growth of invading pathogens. Cease is a preventative fungicide.

Non-Organic Fungicide Control Methods

ALL-IN-ONE: Sold under different brand names, such as BioAdvanced All-in-One Rose or Bonide All-in-One Rose, these products contain different active ingredients in combination (fertilizer, insecticide, and fungicide.) Available as granular, foliar, and liquid. Read the product label to see the active chemical ingredients in each brand. Usually systemic products depending on application.

MANCOZEB: A curative fungicide that is a combination of dithiocarbamates maneb and zineb.

ALIETTE (ALUMINUM TRIS): A systemic preventative fungicide that inhibits spore production, and will not wash off the plant.

WEEDS ARE AWFUL

The easiest way to control weeds is to stop them before they start. Like pests and diseases, early intervention is the key. Spreading a thick layer of compost or compost plus mulch, or laying black landscape fabric beneath your plants, can be effective organic solutions for weed control as they keep the soil shaded, and prevent light from germinating weed seeds.

But be mindful if you choose to use black fabric. While it can be an amazing tool for weed prevention, in hot growing regions, black fabric warms the soil and provides radiant heat on the rose foliage and can cause burning on the canopy. Simply put, the fabric can make conditions for the plant too hot.

Flame weeding is another organic method that can be done early in the season when rose plants are shorter, as it can be more difficult to maneuver the flame wand between taller plants. This method works best when weeds have just emerged (larger weeds are not a match for flame weeding).

You can also solarize the bed with a tarp, if you are preparing a new area to plant roses for the first time, which is a great method for killing weeds before they germinate. The heat from the sun is trapped under the tarp and will kill weeds and seeds within the top six inches of soil. If done properly during initial bed preparation, you will have minimal-to-no weeds in your rose bed.

You can also apply herbicides, both pre-emergent (to prevent weeds before they emerge) or post-emergent (to kill weeds after they have germinated). A common pre-emergent sold under the trade name Preen (Trifluralin) can be sprinkled on the ground around the roses early in the season to prevent the weed seeds from growing. It's safe to use, and will not damage plants if applied correctly. Post-emergents with trade names like Round-Up (Glyphosate) or Paraquat can be used, but do so with extreme caution. If any spray touches the rose it will kill the leaves and possibly the whole plant.

I control 99 percent of the weeds on my farm by removing weeds as they appear. I remove them by hand or with a hoe before they get established. We hoe and remove weeds daily from our beds. Regardless of the weed control method you choose, I leave you with one piece of advice: Get them early when they are small, your future self will thank you.

WATERING & IRRIGATION

The number-one most cost-effective form of care you can give your roses is to water them regularly. People are always looking for fertilizers or other quick fixes to improve their roses, but in reality, all they really need to do is dial in their irrigation program. Roses need enough water to extract the amount they need from soil, but not too much to become water-logged or have "wet feet." If roses do not receive enough water they will be smaller in stature, have fewer petal counts, stunted growth, shorter stem length, decreased yields, be more susceptible to diseases and insects, and their cycles between blooms will be fewer.

> Most rose growers are looking for a magic formula, like "water for one hour each week," but I'm afraid it isn't that simple. The amount of water each rose needs varies wildly by soil type and weather.

Most rose growers are looking for a magic formula, like "water for one hour each week," but I'm afraid it isn't that simple. The amount of water each rose needs varies wildly by soil type and weather. On my 80-acre farm, I have four different soil types between my rose fields and my French Prune orchards. The prunes in the northwest corner block require less water than the block in the southeast corner of the property, because their soil types are different clay and loam percentages. The weather varies subtly in both locations too.

To manage my water and irrigation for my roses, I use more advanced technology. I have a professionally installed soil tesiometer that measures soil moisture every 15 minutes at depths of 12-, 18-, and 24-inches, and sends the data reading to an app on my phone. I realize that this level of technology is not practical for a residential rose garden, or even a small flower farm, but what is applicable are the basic principles of how a plant draws water from the soil or soil tension.

With the changes to the climate in the last 10-20 years, the old adage of "water deeply once a week" simply doesn't work for most zones, especially in the summer months. Stressed roses do not perform well, and cutting roses need to be in top form to have a successful life out of the vase. We need a better way to know when and how to water our roses.

> **KNOW YOUR SOIL**
>
> Soil is comprised of three main components: clay, silt, and sand. Soils are classified by the percentages of these components. You can map your soil at your exact location online. You also can, as I've mentioned many times in this book, get a soil test to determine your soil composition. Knowing your soil texture is useful in determining the water-holding capacity of your soil, as well as how long and frequently to irrigate, and what rate of water you should apply.
>
> **CLAY-BASED SOILS** tend to have high water-holding capacities, but have the lowest percentage of plant-available water.
>
> **SANDY SOILS** tend to have the lowest water-holding capacities, but have the highest percentages of water availability.

How Water Moves through Soil

Water moves through soil by two forces, capillary movement and gravitational movement. Capillary movement is driven by the attraction of water molecules to soil particles and travels through the soil in all directions, but only continues until a soil's moisture capacity is reached. Think of capillary movement like water wicking into a dry sponge.

Gravitational movement takes over once soil capacity through capillary movement has been reached. This is when water is drawn down by the force of gravity. If you continue to add water to the sponge it will eventually start dripping out of the sponge. This is soil saturation. When you don't water enough you may achieve capillary water movement, but not gravitational water movement into the root zone where the rose needs it.

Plant Biology Refresher

First, a little lesson on soil and plant biology. Photosynthesis is the process by which plants capture energy from the sun and convert it into stored energy in the form of glucose. The necessary components of photosynthesis are air (CO_2), water (drawn from the soil by way of the roots), and solar radiation. Transpiration is the mechanism for moving water and minerals from the soil into plant tissues.

The stomata are organs on the leaves made up of two cells that open and close depending on the time of day and environmental stresses. Open stomata allow water vapor and O2 to leave the plant (by-products of photosynthesis), and CO2 to enter the plant (a necessary component of photosynthesis). Open stomata allow for water movement from the soil into the plant. When stomata close, that water movement stops.

What is Soil Tension?

As water is lost at the stomatal openings of the plant's leaves through evaporation, a negative pressure gradient occurs (think of the way a straw pulls liquid), and water from the soil is pulled into the roots and throughout the plant tissue. The negative pressure at the leaves needs to be greater than in the roots in order to draw the water molucules away from the soil particles.

Now, stay with me because I am going to go a little more "science nerd" as I am passionate about helping rose growers understand the role proper irrigation plays in plant health. Soil tension is the force or suction that soil particles exert on water molecules (like the straw analogy I mentioned above). To remove water from the soil, plants must exert greater negative pressure or suction than the suction the soil particles have on the water molecules.

When saturated, the water molecules furthest away from the soil particles are held less tightly by the soil, and easy to remove from the soil. As the soil water is depleted, it becomes increasingly harder for the plant to suck the remaining moisture as the molecules are held more closely and tightly by the soil.

If water is not replenished in the soil, this leads to drought-stressed roses as the roses no longer have available water in their root zone. You may not see this stress right away, but telltale signs come days later with wilting or crispy leaves.

This stress is especially common if the plant is bouncing from extremes over the course of a week. Soil tension is measured as a negative pressure, and the device used to measure it is called a tensiometer. The tensiometer indicates the availability of water in the soil to the plant. In order to get an accurate reading of the water available to your roses, you can install the meter along the water line in the rose's root, and use that measurement to guide your watering and irrigation.

While installing a soil tensiometer will give you the most accurate reading of the water available to your roses, you can also follow a few simple steps to learn how to measure your soil moisture like a human tensiometer. First, water deeply so it looks moist when you insert a spade down to 12-inches deep along the root zone of the rose. Then, do a quick hand test by clumping the soil in your hand. This is where you become a "human" soil tensiometer.

If the soil clumps and holds together in your hand it has enough water. If it crumbles like cake you need to water more. If it is dripping like a wet rag when squeezed, then it has too much water. The goal is to keep the roses at a consistent and comfortable stage of water availability with the soil clumping together so the plant doesn't experience the peaks and valleys of drought stress.

Do a hand-moisture test at intervals of one, three, five, and seven days during different seasons of the year. Record this information in your garden journal, and you will eventually notice a routine taking shape that tells you when and how

> If the soil clumps and holds together in your hand it has enough water. If it crumbles like cake you need to water more. If it is dripping like a wet rag when squeezed, then it has too much water.

long to water your roses. You'll find that in the summer months you likely aren't watering enough. If you're in a warmer climate, you even may need to water during the winter, when seasonal rainfall isn't present.

In locations with more humidity and rain, you may find you need to water less frequently. Typically, a light sprinkling for a few minutes once a day will do no good. The key is longer watering to maintain a consistent moisture level—think capillary vs. gravitational movement until the sponge is dripping. (See the Resource Section for a hand test chart by soil type to help guide you.)

As I've illustrated in this section, there is no magic formula for watering. How frequently you water needs to be customized, as it all depends on your soil type, water output, and weather. What I do know from my years advising members of my learning academy, is that everyone feels better with some general guidelines as a reference to know that they are on the right track (You all really want that magic formula!). Using the concept of soil tension, along with some measuring and good recordkeeping, you will become a master irrigator in no time. Here are some basic water-holding capacities and application rates to guide you as you fine-tune your own watering regimen:

One-inch of water in sandy soils should wet the soil a depth of about 12 inches; in loams, six-to-10 inches; in clay soils, four-to-five inches.

To thoroughly wet a 10' x 10' area two-feet deep requires 125 gallons of water for sandy soils, 190 gallons for loam soils, and about 330 gallons for clay soils.

Once your soil is thoroughly wet, or reached optimum soil moisture, most gardens perform well with a continued application of four-to-five gallons of water per rose per week

> **TAKE SPECIAL CARE WITH POTS**
>
> Roses in pots dry out much quicker than in-ground planted roses because they require more frequent watering. Do your hand water test for moisture in the first two inches of potting soil (instead of 12 inches). If that area is dry, apply a nice soaking of water.

Methods of Watering

Now that you know the goal you want to achieve with watering, you can determine how you want to water. There are so many great options, including handwatering, drip-irrigation in flexible lines of black tubing, and permanent sprinkler installations in a system of underground pipes. On the farm, I use plastic drip-line tubing with 12-inch spacing. Electronic timers can be programmed to ensure your roses receive the water they need without you remembering to do it yourself. If you need help with setting up an irrigation system that suits your property or farm, contact a local nursery, agricultural irrigation supply store, or landscape company to customize a plan.

Your Watering Plan

So how can you put all of this together in a plan for your own rose garden? Here are my four keys for proper irrigation:

STEP ONE: Know your irrigation system's output.

STEP TWO: Know your soil type and soil-holding capacity.

STEP THREE: Measure or gauge pre- and post-irrigation soil moisture levels.

STEP FOUR: Invest in a soil tensiometer (optional).

PRUNING

Pruning garden roses can often send even experienced gardeners into high anxiety. Does this sound like you? Well, I'm here to tell you that pruning is really very simple and hopefully, after this section, pruning will become one of your favorite tasks as a rose grower. My pruning classes are by far the most popular workshops I host at the farm every year.

In-Season Pruning

In-season pruning is often referred to as deadheading, or removing the dead or spent rose blooms. This ongoing task keeps your plant looking attractive and encourages roses to bloom again. The more frequently you deadhead, the more repeat blooms you will have. This is the single, most important task you can do to ensure you have roses to cut throughout the entire growing season. The roses will bloom again eventually even if you don't deadhead, but it will not happen frequently enough to provide bouquets of blooms all year long.

Each rose has sets/groups of leaves, usually in threes or fives, but sometimes as many as seven or more. Together, the set of leaves is called a leaflet. To deadhead

Prune roses about 1/8-inch above the fifth leaflet or higher, selecting a thick, strong cane.

a mature rose you need to cut the stem back to five or greater healthy leaflets. Make your cut about 1/8- to 1/4-inch above the leaf axil. It really doesn't matter whether the leaf is inward- or outward-facing, as a new shoot will emerge from that leaf axil. You may have heard advice about cutting at the outward-facing bud, but those rules were developed for exhibition or show roses. If you don't plan on entering any competitions, then simply cut back a spent bloom to a leaflet on the existing cane, taking into consideration the overall plant shape. When I deadhead and harvest, I don't look for inward- or outward-facing bud eyes, I just cut.

To stimulate new growth and produce a strong new rose, choose a cane that is roughly as thick as a pencil. Some varieties have naturally thinner canes so observe the overall cane size on the plant to determine a good average cane thickness for suitable cutting on each rose. The circumference of the new shoot will grow only as thick as the circumference of the cane from where it's cut. A good rule of thumb is finding a five or greater leaflet that is three-to-five leaf sets below the dead bloom, where the cane is thick enough to grow a new strong cuttable stem.

If you are deadheading a young plant (less than one-year-old), or one that is struggling or isn't needed for cut-flower lengths, the above rules don't apply. Do make sure enough of the plant stays intact so its leaves and canes continue to grow. It's okay to just deadhead to the first leaflet of any size, regardless of cane circumference. Make sure never to deadhead or take a cane down to less than half of the existing height of the plant.

Continue deadheading throughout the growing season until about six weeks before your first expected frost. Then let your rose set hips, aka seed, and naturally proceed in the slow march to seasonal dormancy.

Mid-Season Shape-Pruning or Grooming

By mid-season, roses have completed one or more flushes, depending on your climate, and they may have canes growing wild with abandon.

Here at the farm, by the Fourth of July, I have roses in my field encroaching into my equipment, making harvesting a challenge. Around my house, there is always one cane flopping into my driveway and scratching my car as I pull out of the garage. While I use the Fourth of July as a general guide for beginning in-season pruning, you can shape prune or groom your roses anytime during the season.

So what is shape pruning? Shape pruning is just what the phrase suggests—you are pruning your established roses to the shape that you want them to be. Go ahead and grab the clippers! Trim the cane overtaking a nearby perennial plant, or one that's two-feet taller than the rest of plant creating an imbalance or an eyesore.

Think of this more as trimming the bangs rather than getting a completely new hairstyle. There are no rules for how to shape prune, so use your intuition and prune the canes the way that you like. But remember, these are maintenance trims, so don't cut more than half of the height or half of the total volume of the plant.

> **GROOMING TIP**
>
> When you look back at your garden journal and you find yourself always grooming a rose every year at the same time, it may be time to move that rose to a new location. You may have planted a tall rose in a short spot. Think about moving it to a location with space to grow and thrive to the height it loves to be.

Fall Season Shape-Pruning

In the fall and winter months, wind, rain, and stormy weather return. High winds can damage roses and their canes can snap off, or thrash against each other. Or, if its too top-heavy, it can act as a sail to pull a rose and its roots out of the ground.

In late summer or early fall (depending on your zone), prune to reduce the height of your rose plant and prevent cane breakage. Not all roses require shape-pruning, but this is especially helpful for those tall hybrid teas that may have reached upwards of seven-to-eight-feet tall. A good rule of thumb is to reduce the plant to a height between your waist and shoulders. Use your grower's common sense here, too, and just lop off anything that looks like it will be damaged in rough weather.

Why Prune Roses?

Pruning is like a nice spa treatment for your roses. It sloughs off all the old, dead, and not-so-great parts, and leaves the rose invigorated and ready for a subsequent season of beautiful new growth. The basic principles discussed in this section are for use on hybrid teas, floribundas, grandifloras, and English roses that I grow. My simple methods work for most any variety you're likely to grow.

Shape-pruning is done throughout the season for aesthetic reasons, but is also important to reduce the potential of wind or weather damage in winter.

YEAR-ROUND ROSE CARE

dormant pruning

Follow my rose-pruning method with the five easy steps outlined below, and you will gain the confidence to tackle even the most challenging roses.

BEFORE YOU BEGIN

Think about what and why you are growing each rose. As a guideline, leave landscape roses at three-feet tall and field-grown roses for cut flower production at 12-18-inches tall.

Aim to leave five-to-seven good canes on the plant that are uniformly spaced. Leave the rose structure in the shape of a bowl, with the area in the middle of the plant open and free of canes.

When in doubt cut less off, as you can always reassess and cut more later. Sometimes it's good to sleep on it overnight.

Aim to do your dormant pruning before your buds start to swell and push open for the season. While you can prune later if needed, it's most effective (and easier) to do this before there is new growth.

DON'T BE AFRAID TO CUT! I promise: New canes will grow back. New canes are called basal breaks and their presence is a sign your rose is healthy and ready to keep producing beautiful blooms. Most people aren't aggressive enough when pruning their roses. So go ahead, get in there and make those cuts!

STEP ONE: DOWN

Cut down the rose to a manageable height, but no more than half of the original height, to get a better assessment of how to prune. You can even do this with a hedge trimmer. Cutting to a three-foot height is a good rule of thumb to take down the tall canes so you can see what you're working with.

STEP TWO: THE BIG THREE

DEAD + DISEASED + DAMAGED: Remove any canes that are dead or show signs of disease. Cut back any damaged parts to healthy green canes.

ZONE	PRUNING TIME
Zones 9-11	December - February
Zone 8	Mid-late February
Zone 7	Early-mid March
Zone 6	Mid-late March
Zone 4-5	Early-mid April
Zone 3	Mid-April - Early May

STEP THREE: DINKY
Remove any canes smaller than the circumference of a pencil. It won't support the weight of the bloom and is likely to die in the winter weather.

STEP FOUR: DIAGONAL
Remove canes to prevent cross rubbing and increase airflow to the middle of the plant.

STEP FIVE: DEFOLIATE
Remove any remaining leaves on the plant and clean the bed of all leaves and plant material. This is an essential step in dormant pruning to prevent disease and insects from overwintering on your roses. You don't want to leave any debris or leaves on or around the rose. Place debris material in the trash; do not compost.

AFTER DORMANT PRUNING

After you finish your dormant pruning, proceed to your dormant spray program. Ideally, you are applying your spray within a week of dormant pruning. My favorite dormant spray program is simple – apply a horticultural oil and copper spray. Read the labels on the products for application rates. This all organic one-two combo punch will help prevent both diseases and insects from taking hold of your roses the next season. Aim to get your spray on before buds push out. This winter spray is the step in your rose care program you don't want to miss. Trust me, I skipped a section in my rose field one season and paid the price with familes of aphids and thrips taking up residence in the spring.

Winter Protection

How do you know if your roses need winter protection? In general, hybrid teas, grandifloras, floribundas, and climbers require basic winter protection if temperatures regularly reach below 32 degrees F. In areas that are below 10 degrees F, you will need extra winter protection. The key word here is "regularly." If you have a few nights a year where the temperature dips in the 32-20 degrees F range, there is no need to rush out and winterize your roses.

I experience at least 10-15 days of nightly temperatures below freezing each winter. The temperature bounces up again during the day, so there is no need for me to protect my plants. I've never lost a single rose plant to winter frost. Roses are very hardy and they typically will survive frost and low night temperatures. Winter protection is only needed for growers with sustained cold, freezing temperatures, and/or frozen ground during the daytime throughout the winter. While there are exceptions to these guidelines, in general, most growers in Zones 8-11 don't need to winterize their roses.

Basic Protection

If you live in an area where the winter can be mild, but the ground still freezes regularly, take the steps outlined below.

After you have performed fall shape-pruning and the first fall frost of the year has arrived, prepare roses for a winter nap by insulating the plants with fresh topsoil or a compost "cocoon." (Don't use native soil from around the plant.) This soil cocoon will protect your roses from the roller coaster cycles of freezing and thawing that occur throughout the winter, while maintaining a cozy insulation blanket around the base of the plant and the canes.

Pile the fresh topsoil or compost to cover the plant from where the canes extend from the soil up to one-foot-high. You can also apply dry, shredded, leaves, wood chips or other mulch for even more protection. In the spring, remove the soil mound from the rose, and scatter it around the base of the plant in the root zone.

EXTRA PROTECTION: If you live in an area that experiences sub-zero temperatures, you will need need to create a safe place for your roses to survive during the winter.

After the first fall frost of the year has arrived, prepare roses for winter by insulating them with a mounding "burrito." Cut the canes to three-feet high. Tie them together with twine. Apply at least one-foot of top soil or compost to cover the base of the rose up the canes. (You can add more top soil if you'd like, but do not use native soil.) Liberally cover the soil mound and tied rose with an insulating material like straw, leaves or other non-compacting organic material.

Wrap the mound with wire mesh like cattle fencing or chicken wire as if you are folding a tortilla roll. Make sure the soil and straw are contained by the wire mesh as the burrito's "filling." Be sure whatever wire material you are using has openings for ventilation, and is weighed down or secured to the ground with landscape staples or stakes to ensure that it stays intact. In the spring, you can remove the soil mound and straw burrito from the rose and scatter the material around the base of the plant as mulch.

Develop Your Own Best Practices

Growing Wonder isn't just about following a how-to checklist of rose care (while my Type A personality finds those lists incredibly useful!). This book is also about finding your personal philosophy to plant rearing, and learning to trust in your wisdom even when there are setbacks.

I, too, experience mornings when I walk the rose field to make harvest estimates and count stems only to find a nasty little insect infestation. So, as you wrap your head around the type of rose-care program that fits your lifestyle, I want you to leave this chapter with one important messge: Don't let perfect be the enemy of good. It's easy to be defeated by insects and diseases. There will be ups and downs, triumphs and disappointments.

I've given you the resources to stock your rose tool kit to tackle anything that comes your way. Think of me as your rose-growing personal trainer, and your rose coach who whispers encouragements in your ear every day. It's easy to compare your rose farm or garden to others, but don't do it. You will regularly modify and change your plan, and as nature evolves, you too will evolve your cultural and care practices to grow your BEST roses.

Combine this general guide with your garden journal to customize a care plan that is perfect for you!

monthly care chart

Depending on your location and microclimate, the tasks listed here may shift from one month to the next. By keeping good records during the course of a 12-month period, you will develop a care guide tailored to your roses.

	JANUARY	FEBRUARY	MARCH	APRIL	MAY	JUNE
ZONES 3-5	Purchase bare root roses. Enjoy your winter's rest.	Purchase bare root roses. Enjoy your winter's rest.	Purchase bare root roses. Enjoy your winter's rest.	Purchase and plant bare root roses. Dormant prune rose bushes to 1/2 original height and remove all foliage. Spray dormant oil and fungicide after pruning. Rake and clear debris, leaves, and petals and throw in garbage (Do not compost). Apply compost or compost plus mulch to existing and newly planted roses. Water roses, if no rain.	Dormant prune rose bushes before buds start to swell and remove all foliage. Spray dormant oil and fungicide after pruning. Rake and clear debris, leaves, and petals and throw in garbage (Do not compost). Plant bare root roses. Apply compost or compost plus mulch to existing and newly planted roses. Water roses, if no rain.	Apply a granular slow-release fertilizer or an organic fertilizer at the first signs of leaves. Purchase and plant potted roses. Water roses, if no rain. Monitor and treat for mildew and fungus as needed, if you have summer rains. The first of the year weeds are the worst. Monitor weed growth and remove weeds by hand or hoeing.
ZONES 6-7	Purchase bare root roses. Enjoy your winter's rest.	Purchase bare root roses. Enjoy your winter's rest.	Purchase and plant bare root roses. Remove poor performers. Dormant prune rose bushes to 1/2 original height and remove all foliage. Spray dormant oil and fungicide after pruning. Rake and clear debris, leaves, and petals and throw in garbage (Do not compost). Apply compost or compost plus mulch to existing and newly planted roses. Water roses, if no rain.	Dormant prune rose bushes to 1/2 original height and remove all foliage. Spray dormant oil and fungicide after pruning. Rake and clear debris, leaves, and petals and throw in garbage (Do not compost). Apply compost or compost plus mulch to existing and newly planted roses. Water roses, if no rain.	Apply a granular slow-release fertilizer or an organic fertilizer at the first signs of leaves. Purchase and plant potted roses. Water roses, if no rain. Monitor and treat for mildew and fungus as needed, if you have spring rains. The first of the year weeds are the worst. Monitor weed growth and remove weeds by hand or hoeing. Keep roses watered; set drip system timers and test emitters.	Purchase and plant potted roses. Apply insecticide or introduce predatory insects, if needed. Monitor and treat for mildew and fungus as needed, if you have summer rains. In the morning, spray foliage with water a few times per week to prevent insect populations from growing (omit in humid locations if leaves don't dry out).
ZONES 8-11	Purchase and plant bare root roses. Remove poor performers. Dormant prune rose bushes to 1/2 original height and remove all foliage. Spray dormant oil and fungicide after pruning. Rake and clear debris, leaves, and petals and throw in garbage (Do not compost). Apply compost or compost plus mulch to existing and newly planted roses. Water roses, if no rain.	Purchase and plant bare root roses. Dormant prune rose bushes before buds start to swell and remove all foliage. Spray dormant oil and fungicide after pruning. Rake and clear debris, leaves, and petals and throw in garbage (Do not compost). Apply compost or compost plus mulch to existing and newly planted roses. Water roses, if no rain.	Apply a granular slow-release fertilizer or an organic fertilizer at the first signs of leaves. Water roses, if no rain. Monitor and treat for mildew and fungus as needed, if you have spring rains. Apply insecticide or introduce predatory insects, if needed. Monitor weed growth and remove weeds by hand or hoeing.	Purchase and plant potted roses. Apply insecticide or introduce predatory insects, if needed. Monitor and treat for mildew and fungus as needed, if you have spring rains. In the morning, spray foliage with water a few times per week to prevent insect populations from growing (omit in humid locations if leaves don't dry out). Keep roses watered; set drip system timers and test emitters.	Purchase and plant potted roses. Apply insecticide or introduce predatory insects, if needed. In the morning, spray foliage with water a few times per week to prevent insect populations from growing. Continue to water roses as the temperatures heat up. Deadhead and remove old blooms.	Apply another round of slow-release organic fertilizer. In the morning, spray foliage with water a few times per week to prevent insect populations from growing. Apply insecticide or introduce predatory insects, if needed. Deadhead and remove old blooms.

	JULY	AUGUST	SEPTEMBER	OCTOBER	NOVEMBER	DECEMBER
ZONES 3-5	Purchase and plant potted roses. Apply insecticide or introduce predatory insects, if needed. Monitor and treat for mildew and fungus as needed, if you have summer rains. In the morning, spray foliage with water a few times per week to prevent insect populations from growing (omit in humid locations if leaves don't dry out). Keep roses watered; set drip system timers and test emitters.	Apply the last round of slow-release organic fertilizer. Continue to apply insecticide or introduce predatory insects, if needed. Monitor and treat for mildew and fungus as needed, if you have summer rains. In the morning, spray foliage with water a few times per week to prevent insect populations from growing (omit in humid locations if leaves don't dry out). Continue deadheading to encourage blooms.	Purchase and plant potted roses. Apply insecticide or introduce predatory insects, if needed. Enjoy the last of the season blooms. Stop deadheading and let the roses start to set hips. Shape-prune roses before winter; cut back tall canes that could break or hit each other in storms; cut to waist height.	Water roses, if no rain. Shape-prune roses before winter; cut back tall canes that could break or hit each other in storms; cut to waist height. Start winter pruning, clean beds, and remove foliage. Remove poor performers. Apply winter rose protection. Roses are ready for their winter rest.	Apply winter rose protection. The roses are resting and so are you. Start shopping for new roses for next season.	The roses are resting and so are you. Start shopping for new roses for next season.
ZONES 6-7	Apply insecticide or introduce predatory insects, if needed. In the morning, spray foliage with water a few times per week to prevent insect populations from growing (omit in humid locations if leaves don't dry out). Keep roses watered; set drip system timers and test emitters. Deadhead and remove old blooms. Shape-prune roses, as needed.	Apply another round of slow-release organic fertilizer. In the morning, spray foliage with water a few times per week to prevent insect populations from growing (omit in humid locations if leaves don't dry out). Apply insecticide or introduce predatory insects, if needed. Continue to deadhead and remove old blooms.	Monitor irrigation as needed to make sure roses receive enough water. Continue deadheading to encourage bloom production. In the morning, spray foliage with water a few times per week to prevent insect populations from growing (omit in humid locations if leaves don't dry out). Shape-prune roses before winter; cut back tall canes that could break or hit each other in storms; cut to waist height. Purchase and plant potted roses.	Water roses, if no rain. Monitor and treat for mildew and fungus as needed when rain begins. Enjoy the last of the season blooms. Stop deadheading and let the roses start to set hips.	Water roses, if no rain. Roses are ready for their winter rest. Start preparing for winter pruning, clean beds, and remove foliage. Apply winter rose protection. Start shopping for new roses for next season.	The roses are resting and so are you. Start shopping for new roses for next season.
ZONES 8-11	Monitor to make sure your roses receive enough water and increase irrigation if necessary. Continue to deadhead and remove old blooms to encourage new bloom production. In the morning, spray foliage with water a few times per week to prevent insect populations from growing. Shape-prune roses, as needed.	Continue to deadhead and remove old blooms to encourage new bloom production. In the morning, spray foliage with water a few times per week to prevent insect populations from growing. Continue to water roses as the temperatures heat up. Continue to deadhead and remove old blooms to encourage new bloom production.	Apply another round of slow-release organic fertilizer. Monitor to make sure your roses receive enough water and increase irrigation if necessary. Continue to deadhead and remove old blooms. In the morning, spray foliage with water a few times per week to prevent insect populations from growing. Purchase and plant potted roses.	Water roses, if no rain. Monitor and treat for mildew and fungus as needed when rain begins. Continue to deadhead and remove old blooms. Shape-prune roses before winter; cut back tall canes that could break or hit each other in storms; cut to waist height. Purchase and plant potted roses.	Water roses, if no rain. Shape-prune roses before winter; cut back tall canes that could break or hit each other in storms; cut to waist height. Stop deadheading and let the roses start to set hips.	Roses are ready for their brief winter rest. Start winter pruning, clean beds, and remove foliage. Remove poor performers. Start shopping for new roses for next season.

6

IN THIS CHAPTER

GARDEN ROSES
VS. GREENHOUSE ROSES

PROPERTIES OF A GOOD
CUT-GARDEN ROSE

ROSE WISDOM

SANITATION

THE ROSE HARVEST

PRESERVING
YOUR ROSES

DESIGNING WITH ROSES

HARVEST
& POST-HARVEST CARE

Early in the morning, just before the sun rises, is my favorite time of day on the farm. I grab my "pick list" of the stems we need to cut, take my mug of freshly brewed coffee, and head out to the rose fields. The sweet fragrance wafts over me as I appreciate the beauty I am lucky enough to nurture. I treasure the special mornings, too, when my two boys tag along with me, clipping stems just as I did with my mother and grandmother. It's a full-circle moment for me as I walk my journey as a legacy farmer, following in the footsteps of earlier generations.

I am always inspired by messages I receive through social media or in conversations with members of my learning academy about how enjoyable it is to grow food or flowers, whether one is alone or with others. Roses somehow have an unmistakable power to fulfill just what you need from them, when you need it.

The act of harvesting is a prideful symbol of the special partnership between you and Mother Nature. It's the beginning of a rose plant's final act, coming after weeks during which you have nurtured your baby buds to full-grown rose stems. Harvesting is the grand crescendo, if you will, of your adventure of growing cut-garden roses.

Before you start cutting though, you need a harvest plan. As I mention earlier, I couldn't find a book or online resource that magically answered my questions about how to harvest garden roses for commercial cut-flower production. Most garden rose references devote one obligatory page to tips like cut on an angle, place the stem into water, and enjoy your blooms. That really didn't give me the details I was looking for!

I love to stop and smell the roses while harvesting, especially in this prolific area where my 'Distant Drums' roses flourish.

PHOTOGRAPHY BY JILL CARMEL

HARVEST & POST-HARVEST

I read in early crowd-sourced, flower farming forums that you wait until the bud "pops" before cutting, but what did that mean? That advice was contrary to what I learned in my college horticultural classes on greenhouse rose harvesting.

Bud-popping can look very different depending on garden-rose varieties, nor could I find a simple, easy-to-use formula to follow for harvest and post-harvest care. I wanted to know the best practices so I could produce the best quality of field-grown garden roses.

I sought answers to questions like:

> Can you harvest with a tighter bud—one that hasn't "popped"?
>
> What has the biggest affect on vase life—water, temperature, or something else? (Spoiler alert: it's temperature, not hydration.)
>
> Do I need to use a floral hydration formula, and if so, what kind?

So I went back into research mode, and conferred with commercial greenhouse rose growers. I read research papers on post-harvest care for greenhouse roses, and also explored cold-chain systems in the floral supply industry. From there, I developed my own system for post-harvest care for field- and garden-grown roses.

To put it simply, harvesting when the rose bud is tight is ideal. It's like waiting for the show to start.

If you follow all the steps I cover, your roses will have maximum possible vase life. If you don't have time, you can certainly omit some of these steps and still enjoy a beautiful bouquet of roses, although perhaps for a shorter amount of time.

Roses like to perform! Like any good performer, when the show's over, the makeup comes off and they dash out the stage door. Roses are at their peak performance when they have beautiful, open buds, with petals unfurled, and with pollen and stamens all aglow. Yet, if you clip the flower when it's at full bloom, you've already missed the show.

You'll put it in a vase and the petals fall like quickly. To put it simply, harvesting when the rose bud is tight is ideal. It's like waiting for the show to start, and this method ensures maximum vase life. You won't miss the performance in the vase, and you'll likely enjoy an encore.

My basket is filled with the in-demand 'Koko Loko' rose, harvested at the right stage where the blooms are just beginning to open.

QUICK LOOK HARVEST

YOUR ROSE PERSONALITY

WEEKEND WARRIOR

Harvest at the right stage to maximize the vase life of your roses. Get out there early in the morning, before the temperatures rise, and check those sepals. Strip the foliage off of fresh-cut rose stems, and place in water as soon as possible. **PRO TIP:** Try Quick Dip! Remember, a fully-bloomed rose on the plant will be a sad-looking drooper in the vase.

EVERYDAY GARDENER

For lasting bouquets, don't skip the hydrating and cooling steps. Put cut roses in the refrigerator or better yet, dedicate a beverage-sized or second refrigerator to storing cut roses for at least four hours before arranging.

ASPIRING ROSARIAN

Stick to my entire post-harvest process from start to finish. Harvest at the right temperature and stage. Invest in professional floral hydrator to add days to the life of your cut rose. Consider investing in a walk-in cooler or other refrigerator to keep your roses cooled during the entire post-harvest process.

HARVEST & POST-HARVEST

GARDEN ROSES VS. GREENHOUSE ROSES

The greenhouse roses found by the dozen at grocery stores or floral wholesalers have an assembly-line uniformity. After harvest, they are graded for stem length, bloom size, stem straightness, and other factors so they are cookie-cutter replicas. They will give you five-to-14 days of post-harvest life (with proper care), and generally have mild to no fragrance. They will last longer than a garden rose, but in my humble opinion, they lack the character, nuance, quirkiness, and fragrance of garden-grown roses; all key factors that make garden roses in a class all their own.

Garden roses possess many more unique characteristics than their greenhouse counterparts. They have bloom sizes that vary widely, depending on cultivar. Some have long, straight stems, while others have dainty, curved, or arching stems. They also have fragrances that will sweep you off your feet. Many of my floral design clients actually prefer roses with slightly curved stems or petite blooms. These florists choose garden roses for their weddings and events, because the fragrance profiles are unmatched by current greenhouse varietals. These traits of garden roses also mean that their vase life is shorter than what you may be used to from a grocery store cooler. With proper harvest and post-harvest care, garden roses will have a vase life of three-to-seven days.

Varietal selections will ultimately affect your vase life and post-harvest performance. I rely on the six main factors (opposite page) when selecting roses to grow as cut flowers; however, not all rose cultivars perform across six categories. I would probably eliminate a rose if it only has one of the six characteristics. The best ones possess all six, such as 'Princesse Charlene de Monaco', 'Golden Celebration', 'Francis Meilland' and 'Grande Dame', which check all of the boxes.

Many hybrid teas have superb vase life, long, straight, thorn-filled stems, and thick, voluptuous petals but only mild fragrance, while many English roses have a fragrance that makes you swoon with delight and pretty, curvy, stems with petals that fall within a day of picking.

You may want a rose that rivals a greenhouse-grown stem with seven days of vase performance, but the trade off is a rose with more thorns. If you plan your cutting garden with a nice assortment of varieties with different characteristics, you'll always have a beautiful selection of roses to cut.

Each of the roses I grow is singularly beautiful, and this vignette reveals just that. The varieties featured here include 'Intrigue' (dark plum); 'Angel Face' (deep lavender); 'Nicole Carole Miller' (mauve) and 'Teasing Georgia' (yellow).

PROPERTIES OF A GOOD CUT-GARDEN ROSE

What makes a garden rose good for cutting? There are six key factors to look for at harvest time.

BUD SIZE & PETAL COUNT
Bigger isn't always better with a garden rose.

SHADE
Intensity and Color of Flower: These characteristics are subjective to the beholder. From bright and bold to soft and hazy tones, they all have their own wow factor.

LEAF COLOR AND SIZE
Deep-green is usually seen as better quality. These variables are also subjective, but give me a bi-color leaf any day.

STEM THICKNESS
It's best to select stems about the diameter of a pencil, or thick enough to support its bloom without drooping or breaking.

STEM LENGTH
Ideally, 12-inches or greater.

FRAGRANCE
The more intense and unique, the better.

ROSE WISDOM

Hybrid teas have the longest vase life, followed closely by floribundas, grandifloras, and English varieties.

Single-petal, garden-rose varieties have a shorter vase life than multi-petaled.

There is no formal grading system for garden roses like there is for greenhouse-grown varieties, but many of the same measures are used to harvest garden roses.

On first- or second-year plants, be careful cutting longer stems, because the same harvest rules that apply to mature plants do not apply to younger ones. As a general rule, don't cut a full stem length (12-inches or greater) on a first-year plant (bare root or potted) until it has at least doubled in size, or usually six-to-seven months after planting.

For some varieties, it's possible to cut late in the season on first-year plants. Time to cutting full stems from planting will depend in your zone. Cutting a stem too long will weaken the plant before it is fully established. On young plants, you can cut shorter stems to display in short bud vases or arranging simple posies.

You can cut at a 45-degree angle if you'd like, as it will help with water uptake, but it's not necessary. During all these years of harvesting, I haven't seen measurable difference in vase life or quality between straight cut vs. angled cuts.

There's no need to immediately submerge stems in water or cut under running water. When recutting just-harvested stems at the processing table, place them in water or in a water-hydration solution. Any air bubbles that enter will have minimal to no effect on the vase life or the quality of a rose, especially if you use a commercial floral hydrating product.

Follow the mixing instructions for the correct hydrator-to-water ratio. Much like fertilizers and pest control measures, if it's not mixed correctly the solution will not improve the holding time or vase life of your rose. Having the correct dosage is crucial.

Thorns: Leave them alone. If you don't need to remove them for comfort—such as for a bridal bouquet that needs to be held—keep thorns on the stems to avoid having multiple open wounds for bacteria to enter the stem tissue, which, in turn, decreases vase life.

The popular range of tawny rose blooms, represented here with bud vases containing 'Koko Loco', 'Honey Dijon', 'Charlotte' and 'Lava Flow'.

SANITATION

This is one of the most frequently overlooked steps during rose harvesting and post-harvest care, but arguably it's the most important. If you start with a dirty vessel—one that most likely contains bacteria and fungi—all of your other post-harvest treatments will be less effective. Studies have shown that a dirty bucket can reduce the vase life of a rose by 20 percent!

To sanitize, clean your vessels, bucket or vase, clippers, floral cooler or refrigerator, tables, countertops, or the processing area with hot, soapy water, a water-bleach mix, or a commercial floral disinfectant. For home gardeners, hot, soapy water or a water-bleach solution will do just fine.

> Studies have shown that a dirty bucket can reduce the vase life of a rose by 20 percent!

Remember, don't mix dish soap or other household cleaners with bleach. Some dish soaps and household cleaners may contain ammonia. When ammonia is mixed with bleach, it forms a toxic gas called chloramine that is released into the air, which has the potential to be fatal if inhaled.

Be sure to check your product labels carefully when mixing cleaning materials with bleach or don't get fancy, just stick to a simple water-and-bleach mix for sanitation.

For Aspiring Rosarians wishing to up the sanitation game, invest in a commercial floral disinfectant like D.C.D from Floralife, or Clean Touch or Professional Cleaner from Chrysal. A comparison study conducted by an independent laboratory found that buckets cleaned with D.C.D cleaner showed a one-count of bacteria, while buckets cleaned with bleach had 339 times the amount of bacteria, and those just rinsed with plain water had a whopping 800 times more bacteria.

For me, as a professional grower, I don't want to be recleaning buckets from the shelf every single day, so I like the residual effect of a professional floral cleaner when cleaning buckets, cooler walls, clippers, and surfaces. For most home growers who sanitize tools and immediately use them, soapy water or a bleach solution are great for day of sanitation before harvest.

THE ROSE HARVEST

Before you begin the harvest process, keep in mind three variables that affect vase life: the temperature during flower storage from harvest to vase; the stage the flower is cut; and light exposure.

As I said earlier, the best time of day to harvest is in the early morning when roses are the most turgid, hydrated from the cool nighttime atmosphere, and ready to pick. Only harvest what you can process (get into water) in a 30-minute period. If you need to keep harvesting, place the roses in a floral cooler or the coolest, darkest location you can until processing. Don't harvest if the temperature is above 80 degrees F, as it will lead to disappointing vase life.

The standard length of a commercial cut garden rose is at least 12-inches, so plan to make your cut just above the first fifth-or-greater leaflet and at least 12-inches from the top of the bloom head, just as you have done while deadheading but farther down the stem. If you need a longer stem for a tall vase, then you can cut longer; likewise, cut a shorter stem for a bud vase. The length you cut is the length you need; just remember: never cut more than one-half of the plant's total height.

Don't focus on the bloom itself and how it's opening; focus on the sepals to tell you when it's ready to harvest. Harvest when at least three-quarters of the sepals are reflexed and the rose is in tight bud-to-marshmallow stage beginning to unfold. All of the sepals collectively are the calyx. Harvesting is a delicate balance. You want the bud with just enough sepals in the calyx reflected, but not open enough to allow pollinators into the bloom.

Once the bloom is pollinated, it sends a signal to the rose to set seed and begin the reproductive cycle. This signal causes a rose to drop its petals, dramatically decreasing the vase life of the rose. This is why many people don't have success harvesting their roses and end up with petals all over the dining table. They harvest

HARVEST TOOLS

Rose gauntlet gloves	Water Bucket, Vase or other Vessel
Floral Sanitizer	Floral Hydrator
Bypass Clippers	Stem strippers (If desired)
Harvest Basket or Tub	

when the bloom is partially or fully open and likely after an insect has pollinated it.

When placed in a vase, this post-harvest pollinated rose may develop bent-neck or shatter often within hours of picking. Be very careful picking even partially open, or "popped" blooms, as pollinators have a way of sneaking in and out of that mountain of petals without you even seeing them. So can you harvest when the bud is open? Yes, but the time it will have in the vase will be less.

If a bloom looks like it's ready to harvest, it probably is. Remember to document this process in your garden journal or take photographs to help you remember whether a rose was harvested too early or at just the right stage, and note how long a bloom lasts in the vase.

After you've cut your roses, place them horizontally in a picking basket or tub. I use everything from wicker baskets to plastic bulb crates, and regularly lay the stems flat in the back of my ATV's flat bed. Really, anything goes to get your roses from the garden to the final processing area. (Just make sure it's sanitized first!) Be careful during transport as petals are delicate and can bruise easily.

Your processing area could be the kitchen table, design studio counters, or work bench in a barn. If you have a walk-in cooler, you can even process inside the cooler. The cooler the area the better. If you are a flower farmer, harvest in an air-conditioned room or a shaded outdoor workspace.

Processing the roses at an indoor worktable is much more efficient for filling orders and maintaining an optimal post-harvest temperature. Just prep your clean buckets or vases with a hydrated solution, remove the rose stem foliage, recut the stem, and place immediately in the water bucket or vase.

Harvest the Menagerie Way

STEP ONE: Gather tools for harvesting. Sanitize and clean all tools and the floral-processing area.

STEP TWO: Dry-harvest stems in the early morning, before temperatures reach above 80 degrees F. The cooler the temperature, the better. Harvest stems at 12-inches or greater with three-quarters of the sepals down, and the flower is in tight bud-to-marshmallow stage.

STEP THREE: Place stems in basket and transport to cool processing area.

STEP FOUR: Remove all foliage below the water line. No need to remove thorns.

HARVEST & POST-HARVEST

STEP FIVE: Bunch rose stems as desired, and tie with rubber bands. Trim the bottom of the stems.

STEP SIX: Place in vessel with water and floral hydrator.

STEP SEVEN: Chill roses in floral cooler or refrigerator at 34 degrees F for a minimum of four hours with an ideal chill/hydration time of 24 hours. They can stay chilled for up to seven days, depending on variety.

STEP EIGHT: Remove roses from cooler when you are ready to design with them!

The bloom phases of 'Golden Celebration' revealed in this collection—from tighter bud (ideal harvest stage) to full bloom at right.

Prepare for Post-Harvest

Most people believe it's crucial to immediately plunge freshly-cut garden roses into a bucket of water, but contrary to popular belief, that will not help you to achieve the maximum vase life or preserve the quality of the bloom.

If it's 85 degrees F outside when you harvest, or if you leave a freshly-picked bucket in a 75 degree F sunny place to hydrate, those blooms will likely have a shorter vase life, drop their petals and generally not be ideal elements of a floral arrangement.

HARVEST & POST-HARVEST

Maintaining the proper temperature at harvest and post-harvest has more of an effect on the overall vase life of cut garden roses than hydration. Simply put: Heat kills flowers.

For every 10-degree temperature increase the speed at which the physiological processes inside the stem increases two-to-three times. Likewise, for every 10 degree temperature decrease, the physiological process is slowed two-to-three times. Maintain the coolest environment you can—from harvest through processing and storage—to increase the vase life, and ultimately preserve the quality of your cut roses.

Post-Harvest Steps

Store roses flat and dry in a cooler or fridge, or place on the floor in a cool, dark closet until you have time to follow my post-harvest steps. It's best to process within a couple of hours or you will see the effects of deterioration on the rose.

Prepare the rose stem to drink on its own. Take the vessel (bucket or vase) you sanitized during your pre-harvest prep and fill it with water and floral hydrator. Follow the directions on the bottle for water-to-hydrator ratio.

Commercial floral hydrating products improve water uptake by the stems, have an antibacterial agent, prevent bent-neck, as well as drastically lower the pH of the water, making the molecules more hydrophilic (a fancy way of saying they stick together better). These solutions have no sugars in them, or use a one-size-fits-all hydrator that's great for Weekend Warriors, such as Quick Dip, discussed in the next section. Your rose stem is like a big straw after it's picked. A system of xylem capillaries move water through the stem carrying nutrients.

After a rose is picked, these little capillaries can become blocked like a thick milkshake stuck in a straw. The blockage can slow the hydration of the stem and lead to decreased vase life. A "rose straw" can also get blocked post-harvest when in a stressed environment with high heat or lack of water. Commercial floral hydrating products keep your straw clear from blockage and your rose hydrated. Adding a floral hydrator to your post-harvest routine will change the game, doubling your vase life, and maintaining the quality of the rose. If you don't have a floral hydrator, you can just place your roses in a vessel filled with plain water.

Now you're ready to get your stems in the water vessel. Remove all leaves that will fall below the waterline, and any outer bloom petals that are damaged or diseased.

Leaves below the waterline can cause bacteria in the water that will clog your "rose straw." You can bunch your stems in a quantity that works for you and wrap with a rubber band or place directly in the vessel unbunched. I bunch, securing roses with two rubber bands in 10 -, 15- or 25-stem bunches.

Bunching before bucketing also helps maintain bloom quality because the stems won't flop in the bucket or get bruised. Make another 1/4-inch cut on the bottom of the stems to remove calluses. This step allows you to even up the stems. Immediately place rose stems in your prepared vessel with water and floral hydrator.

> Remove all leaves that will fall below the waterline, and any outer bloom petals that are damaged or diseased. Leaves below the waterline can cause bacteria in the water that will clog your "rose straw."

Place roses in a 34-38-degree F cooler with a relative humidity of 80-90 percent for a minimum of four hours, although 24 hours is best to give the roses enough time to cool and hydrate. Remember, temperature is the number-one factor influencing the vase life of your garden rose. By lowering the temperature and allowing the flower to hydrate, you are slowing its metabolism (respiration and transpiration). If you don't have a floral cooler or refrigerator, your rose will still last, just not as long. If you follow all of my harvest steps—place your roses in a water-hydrator solution and store in a cool, dark spot—you'll still enjoy more days of vase life.

Once your roses have chilled for 24 hours, you can remove them from the cooler and place them at room temperature to begin the process of opening, aka blooming. You also can leave your roses in the cooler for an additional two-to-three days.

In general, hybrid teas can stay up to one week in a cooler, while English varieties need to quickly get on with the show. Use your grower's instinct and see what roses last longer for you after a holding period in your post-harvest storage conditions.

You've harvested the fruits of your labor, and now it's time to enjoy watching your roses unfurl. Give yourself a pat on the back, too, because you reached the finish line giving your garden roses the care they need. Let the blooms open to the size that you like, and now you are now ready to design with them. The next step in the cut-flower journey is to create something magical. I love the many ways you can style and enjoy garden roses.

PRESERVING YOUR ROSES

So what are floral preservatives, aka flower food, and why should you care about them? Floral preservatives provide a nutritional source for the stem, maintain hydration, and help prevent the growth of bacteria. This useful trifecta increases the longevity of the cut garden rose. Flower foods come in many different options, from hydrators and holding solutions, to feeding preservatives.

Some commercial products include all three elements, and some are standalone, so it's important to look at the description and ingredients on the product label to see what type of flower food it is. In the world of commercial cut-flower production, they are often referred to as step one, step two, or step three products. Step one is hydrator or conditioner, step two is holding solution, and step three is feeder or what most people commonly call flower food.

HYDRATION & CONDITIONING: This type of product will hydrate your roses and provide a bactericide to prevent blockage to the xylem tissues while also lowering the pH of the water. This is step one in the post-harvest process. These solutions have no sugars to aide in feeding the rose.

HOLDING OR TRANSPORT: These solutions keep roses in a holding pattern for transport. This is step two in the process. This product is usually used by commercial growers who need to transport flowers by truck or air over a period of days, such as from their farms to a sales outlet like a grocery store, floral shop, or flower wholesaler. The solution provides a bactericide to keep the water clean, and sugar as a food source for the stem. It's formulated to give cut flowers just the right amount of sugar to keep them alive, but not to hasten the process of opening the buds. Most home growers will not use this step.

FEEDING: This is what most people commonly call flower food, found in little powdered packets that come with a bouquet of flowers from a grocery store or florist. The food contains a combination of sugar as an energy source for the rose to bloom, an acidifier to help increase the uptake of water, and bactericide that keeps stems unplugged and water moving through the tissues. Flower food is sold in powdered packets, liquid packs, or even in commercially-sized jugs for florists. This is step three, or the final step in the post-harvest preservative process. We include a pack of flower food with all of our rose bunches that leave the farm.

HARVEST & POST-HARVEST

post-harvest processing

Adopt good post-harvest care habits, practices, and treatments to ensure the longevity of your roses.

A WORD ABOUT ETHYLENE

Ethylene is a colorless odorless gas circulating in stealth mode around the atmosphere, inside your house or floral cooler, and it can affect the vase life of garden roses. Depending on the rose cultivar, exposure to ethylene can cause a flower to age prematurely, drop leaves, dip buds and have decreased vase life. Ethylene can be a problem for flowers stored for longer periods of time in floral coolers; travel on transport trucks; sit in a bucket on the showroom floor at a wholesaler; or the display case at a grocery store; or even if they are stored for too long in a home refrigerator.

If you are a home gardener or flower farmer, you need to be mindful of ethylene levels. Be aware of the potential exposure to ethylene levels when storing roses for long periods or near fruits and vegetables. For commercial flower farmers or Aspiring Rosarians, you can invest in an ethylene monitor for your floral cooler, or use a product that blocks ethylene.

Garden roses have shorter shelf-lives compared to other cut flowers or greenhouse-grown roses. Most are placed in a cooler for 24-to-48 hours after harvest, and then transported to the end user who's ready to design in a week or less. This is why garden roses, if harvested at the right stage, stored properly and quickly delivered to the end user, do not have time to be aged by ethylene even in a home refrigerator.

Quick Dip is one of the many methods for hydrating and conditioning cut roses.

USING QUICK DIP

This is the perfect floral hydrator for people on the go! Quick Dip is an instant hydrating product that you don't need to mix with water. It aids in the uptake of water to prevent bent-neck, and it keeps stems clear so they can drink more water. It's so simple to use. After cutting the rose stem, dip it into two inches of Quick Dip for one second. Only one second, that's it! Then you place the stem directly in a water bucket. You can use Quick Dip in place of, or in conjunction with, a traditional water floral hydrator solution. Available by the pint, this is a great option for the Weekend Warrior or Everyday Gardener who may not want to invest in a large jug of professional hydration solution. Use it on all flowers, not just garden roses.

COOLING

Garden roses have an extended vase life when stored post-harvest at a temperature of 34-38 degrees F with hydration. For Weekend Warriors, just use the refrigerator you have at home. Everyday Gardeners might consider investing in a second refrigerator that can be placed in the garage and be used just for flowers. Aspiring Rosarians may want to invest in a pre-fabricated, walk-in cooler (new or used), or build your own with insulated walls; an air conditioner and a Cool-Bot temperature device.

Keeping a consistent, cool temperature from harvest to post-harvest is the key to a long vase life for garden roses. Clean regularly with a floral sanitizing solution.

DESIGNING WITH ROSES

I'll be honest, I am not a professionally trained floral designer. I am a farmer and I'm still acquiring my floral design skills everyday. Now, after years of help from many floral designers, even this farmer can take a simple garden rose and create an arrangement! Some days it still feels like I'm all thumbs, but with practice I've gained the confidence to make simple rose bouquets with ease and elegance. So, go easy on yourself as you create your designs and let the natural beauty of the rose guide you. I promise, she won't steer you wrong.

> Go easy on yourself as you create your designs and let the natural beauty of the rose guide you.

GETTING A ROSE TO OPEN QUICKLY: If you didn't have time to plan ahead to allow blooms to open, all is not lost. Place roses with closed buds in warm water with flower food. The water needs to be warmer than room temperature, but not hot. Do the hand test for the temperature the same as you would for giving a child a warm bath. Place the bucket of warm water and roses in a naturally lit location. They should open within a few hours.

REMOVING THE GUARD PETALS: Guard petals are the outermost layer of petals on a rose. They are, as the name suggests, the petals that protect the rose's inner petals, maintaining the bloom's quality and integrity. They may have browning, bruising, discoloration, or insect damage. This is perfectly normal for any rose, especially garden roses that are more prone to bruising and browning. Guard petals are crucial for protecting roses during the harvest, transport, and/or shipping process, so the interior of the rose stays in good condition. Removal of the guard petals is just another step in post-harvest care.

You can do this easy step when the rose is in a closed-or-marshmallow bud stage, or after the bloom has opened a bit more. Place your index finger inside the outer guard petal and gently peel it away from the petal base where it meets the calyx, pulling down toward the stem away to remove the petal. Do this slowly and gently, with care not to damage the remaining petals on the rose. Repeat the process, removing the imperfect outer petals until you have a blemish-free rose. Now you are ready to design!

Stems of 'Princesse Charlene de Monaco' roses practically arrange themselves in a simple vase.

DESIGN STYLE

YOUR ROSE PERSONALITY

WEEKEND WARRIOR

Clip one or more stem each week to display in your favorite bud vase(s). Place in a location that you can enjoy daily, like a bedside table or bathroom counter, and don't forget to stop and smell your rose every day!

EVERYDAY GARDENER

Try your hand at wrapped bouquets to give to family and friends for special occasions or just because. Make it easy on yourself, and skip the greens to make it an all rose bouquet.

ASPIRING ROSARIAN

Practice designing with your garden roses weekly. With frequent practice, you'll learn which varieties open quickly and which ones have a longer vase life. Get to work creating everything from small bud vases to wide compotes. Before you know it, you'll have a go-to formula and style for designing with garden roses in any vase shape and size for any occasion.

PHOTOGRAPHY BY JILL CARMEL

Simple Steps for Designing with Garden Roses

Be sure the vase or vessel you are using has flower food added. While not necessary, it will help the roses last longer, to keep the water flowing, it will reduce the bacteria, and extend the life of your blooms.

Remember to recut the stem, before you place a rose in a new vase. A 1/4-to-1/2-inch fresh cut is good enough, but you can trim more if you see any browning, or if the ends of the stems became mushy.

Don't remove the thorns unless you need to. Removing the thorns adds wounds where bacteria can enter the stems, decreasing vase life. If you're designing in a vase or making a centerpiece, leave the thorns on the stems. For a hand-tied bouquet, wedding bouquet, or other presentation where the rose stems will be held, you can remove them by hand or use a thorn stripper.

Make sure the vase is clean and disinfected before designing in it.

Go foraging in your yard. Find additional elements from your own garden as design companions to your garden roses. Let the roses be the star of the show with your garden's elements as the supporting cast.

Avoid using floral foam with garden roses. It's like a death sentence for them. They'll be wilting almost immediately when you place them in the foam.

Transport your roses in the coolest and darkest space possible. Remember, temperature has the greatest effect on vase life, so if you are taking a wrapped bouquet to a friend's house, be mindful of the temperature and sunlight. Place the flowers in the back of your car on the floor in the shade with the air conditioning vents pointed on the bouquet - don't put it on the hot front seat with the sun blazing through the window.

Keep it simple. I know this sounds cliché, but the most breathtaking arrangements are the ones that are the easiest to make. Don't be afraid to design with a single color family, or even a single variety. Sometimes, it's best to allow your roses to shine all by themselves.

Just keep cutting. You planted a rose-cutting garden to enjoy so don't forget to do the most important part, the harvesting. Even if it's just a stem a day to display beside the kitchen sink, on a bedside table, or vase on your desk at work.

These beautiful 'Princesse Charlene de Monaco' roses have been harvested at the right stage, conditioned, cooled, and bunched -- and are now ready to be arranged in my vases.

HARVEST & POST-HARVEST

simple garden roses in a vase

The most wonderful thing about an all-rose arrangement is that no two look the same, and even when you think it needs something more, it's likely perfect.

BEFORE YOU BEGIN

Collecting and displaying roses from adjacent color families is an easy way to make a lavish display with up to two dozen roses. Keeping it tonal is my favorite way to display roses in my home. When I teach at the farm, I always jokingly say this technique is a very specialized, honed over many years, highly technical "stuff the vase with roses as full as you can method."

I honestly just keep adding more roses until I have no more room in the vase. Then, like Coco Chanel looking in the mirror and removing one accessory before leaving the house to be sure she didn't overdo it, I remove one rose. So use as many as you'd like to fill the vase you have. There are no magical rules or stem counts here. For a more natural look and for long lasting enjoyment, use roses that are at different stages of bloom development, from wide open to more closed bud-to-marshmallow stage.

INGREDIENTS

16 stems of garden roses: 'Golden Celebration', 'Lady Emma Hamilton', 'Distant Drums', 'Carding Mill' and 'Princesse Charlene de Monaco'.

TOOLS

Vase
Clippers
Flower Food
Floral Frog or Pin (optional)
Lazy Susan (optional)

STEP ONE: Place a floral frog or pin in the bottom of the vase if you like to aid in placement of the rose stems (floral adhesive may be needed to secure). Fill the vase with water and floral food.

STEP TWO: Remove guard petals on roses and any foliage that will be below water line.

STEP THREE: Working from the perimeter inward, place the taller stems on the outside edges to give the design its structure. I like to think of it like framing a house.

STEP FOUR: Continue adding stems at different heights and layers to add dimension. The key here is to cut the roses at varying heights and stagger their placement so they aren't like crayons in a box, but like people sitting in the bleachers.

STEP FIVE: Fill in any remaining spaces so the roses look like they are effortlessly cascading out of the top of the vase with no gaping holes.

Use your creative license and determine what looks good to you. The rose stems you use don't need to be stick-straight. My favorites are the arched stems and half-popped rosettes to give arrangements a natural feel that mirrors the environment.

PHOTOGRAPHY BY ASHLEY LIMA

IN CLOSING

A CHORUS OF WONDER

You've reached the crescendo in your journey as you've followed my best rose-growing advice through the steps of planning, planting and caring, to harvesting and designing. Now you get to enjoy the magic of the blooms you have grown and hold them in your own two hands. While this book is peppered with a dose of scientific processes that will help you grow exceptional garden roses, don't forget to trust your own instincts, too.

I encourage you to plant what gives you wonder and joy. Like fashion labels, rather than buying into trends that inevitably end up staying in your closet, neglected, be true to the roses that speak to you. Fill your landscape with the roses and plants that you love, and you will forever enjoy the abundance the garden brings.

When it comes to caring for your roses, don't get stuck over-thinking the process either, or you might forget to enjoy the bounty. Get to know the intimate details of your roses: the fragrance, hips, petal and foliage color, and unique personality of each cultivar. In a garden, they all come together in a chorus of wonder.

I often find myself rediscovering their enchantment after all of these years. There are the ones that take you by surprise, the ones that loom larger than life, and the sublime beauty that captures your heart and becomes your one true rose love. A garden rose is so much more than a list of traits, but the feeling that washes over you when you inhale the deep breath of its intoxicating scent is truly like a warm hug welcoming you home.

I still dawdle through the roses on carefree weekend mornings with my husband and boys, smelling every bloom just as I did as a child with my mother and grandmother. Garden roses are a part of my history: in my childhood throwing petals into the wind with abandon, and as an adult farmer harvesting fresh-cut roses every day. It's likely that you also began your journey with roses many years ago. We all have a special memory of the unforgettable rose—whether it was the stems we picked alongside our grandmothers; the corsage or boutonniere worn for a school dance; the single stem given to you by your first love; or the ruffled petals tucked into a wedding bouquet. These are moments in time that are forever cherished.

My wish for you is that you remain forever dreaming of roses while you nurture the garden of life.

Flowers have followed my family through the generations, from my grandmother and uncle (with friends) in an Oxnard, California, zinnia field in the late 1940s to present day, as I stand with my two sons in the rose fields at Menagerie Farm & Flower.

As you continue in your rose-growing journey, remember that your path may be filled with zigs and zags, circle and squares, as was mine. I challenge you to take the time to write in your journal, noting three things each week that you are grateful for in your rose garden. This mindful practice will help you on the days you may get to a bump in the road and feel defeated by Mother Nature. It is at these times your notes can help you pause and see the beauty you created with her, as you continue tending to roses together.

My wish for you is that you remain forever dreaming of roses while you nurture the garden of life. I want you to remember this: a rose garden is a place for the heart to live, the mind to grow, and the soul to flourish, so make it a space with plenty of room for you to live and enjoy.

resources

Said in my best Julie Andrews voice from *The Sound of Music*, these are a few my favorite things. From tools to reference books, I'm sharing all my most-valued resources with you. I encourage you to shop at your local independent garden centers and nurseries, too.

FROM MY FARM

Menagerie Academy
academy.menagerieflower.com

My learning community provides dynamic resources for flower farmers and garden lovers to learn and grow. Garden rose educational offerings range from a private online membership community to 1:1 consulting sessions, hands-on workshops at the farm and online courses. You're sure to find a place to call home to go beyond the book and continue your garden rose education.

Menagerie Farm & Flower
menagerieflower.com

I've gathered my favorite plants and rose-care products in a one-stop shop.

Menagerie Farm & Flower Amazon
amazon.com/shop/menagerieflower

I know many of you love the convenience and fast shipping that Amazon provides, so I've compiled my must-haves in one easy-to-shop storefront. I receive a commission from Amazon for products purchased through this link.

BOOKS AND COURSES

Perfect for inspiration as you expand your knowledge of the rose, plan your rose garden, or farm design, or start your journey in floral design.

GARDEN DESIGN

Cultivating Garden Style
by Rochelle Greayer

Patina Living
by Brooke and Steve Giannetti

The Complete Gardener
by Monty Don

The Gardens of Bunny Mellon
by Linda Jane Holden

The Layered Garden by Dave Culp

Garden Design Bootcamp
with Rochelle Greayer of Pith + Vigor
pithandvigor.com/classes

FLORAL DESIGN

Cultivated: The Elements of Floral Style by Christin Geall

Floret Farm's A Year in Flowers
by Erin Benzakein

Martha's Flowers
by Martha Stewart with Kevin Sharkey

Seasonal Flower Arranging
by Ariella Chezar

The Art of Wearable Flowers
by Susan McLeary

Heartfelt Floristry
with Gabriela Salazar of La Musa de las Flores www.lamusadelasflores.com/heartfelt-floristry

Poetry of Flowers
with Jen Lagedrost Cavender of Nectar + Bloom
www.nectarandbloomfloral.com/poetry-of-flowers-membership

Susan McLeary Courses
www.susanmcleary.com/courses

ROSE REFERENCE BOOKS

A Rose by Any Name: The Little-Known Lore and Deep-Rooted History of Rose Names by Douglas Brenner and Stephen Scanniello

David Austin's English Roses by David Austin and Michael Marriott

Growing Good Roses, A Year in the Life of a Rose by Rayford Clayton Reddell

Passion for Roses by Peter Beales

The Rose by David Austin

The Rose Bible by Rayford Clayton Reddell

ROSE NURSERIES

There are many wonderful nurseries across the United States and abroad growing wonderful, high-quality plants. It would be impossible to list them all. Here is a list of sources that stock both bare root and potted roses to get you started.

Antique Rose Emporium
antiqueroseemporium.com

David Austin Roses
davidaustinroses.com

Heirloom Roses
heirloomroses.com

Menagerie Farm & Flower
menagerieflower.com

Rogue Valley Roses
roguevalleyroses.com

ROSE DIRECTORIES & ORGANIZATIONS

There are rose societies all around the globe, and I encourage you to connect with a local organization in the place you call home.

American Rose Society
rose.org

Canadian Rose Society
canadianrosesociety.org

Heritage Rose Foundation
heritagerosefoundation.org

National Rose Society of Australia
rose.org.au

The Rose Society UK
therosesociety.org.uk

Help Me Find Roses
helpmefind.com/roses

This is a great place to start your search for all things roses. Join their easy-to-use online message boards to find like-minded rose lovers.

JOURNAL AND NOTE-TAKING

Asana
asana.com

Bullet Journal
bulletjournal.com

Menagerie Academy Journals and Notekeeping
academy.menagerieflower.com/resources

Simplified by Emily Ley
emilyley.com

The Garden Handbook
by Whitney Hawkins
thegardenhandbook.com

FELICIA'S FAVORTE SUPPLIES

Apron: MEEMA Waist Apron Denim

Clippers: ARS Pruning Shears 130DX

Fertilizers Foliar: Neptune's Harvest Rose & Flowering Formula 2-6-4 and E.B. Stone Fish Emulsion w/ Kelp

Fertilizers Granular: E.B. Stone Organics Rose & Flower Food

Dr. Earth Total Advantage Organic Rose & Flower Fertilizer

Flame Weeder: Red Dragon Backpack Flame Weeder

Gopher Basket: Digger Root Guard Gopher Basket 5 gallon

Hats: Outback Trading 1497 River Guide UPF 50 Waterproof Breathable Outdoor Cotton Oilskin and Sloggers Women's Wide Brim Braided Sun Hat with Wind Lanyard Style 442DB01

Loppers: Fiskar 15-inch Power Gear Super Pruner

Mycorrhizae: Wildroot Organic Mycorrhizae and Xtreme Gardening Mykos Pure Mycorrhizal Inoculant

Rose Gauntlet Gloves: Bionic Rose Gardening Gloves

Soil Test Kit: Soil Kit by Soil Test Kit

Sprayers:
Chapin International 21220XP 2-Gallon Premier Pro XP Poly Sprayer

STIHL SR 450 Backpack Sprayer

Sunblock: MDSolarSciences Daily Perfecting Moisturizer SPF 30

Temperature Monitoring: TempStick

Tensiometer: Irrometer Model SR 12-, 18-, or 24-Inch

Water pH Test Kit:
pH Perfect pH Test Kit

FAVORITE HARVEST & POST-HARVEST PRODUCTS

Chrysal
chrysalflowerfood.com
Shop online for my top picks RosePro Hydration and Rose Pro Liquid Flower Food Sticks

Cool Bot
storeitcold.com
Shop for build-your-own and prefabricated cooling units.

Eco Fresh Bouquet
ecofreshbouquet.com
Biodegradable hydration wraps for transporting your garden bouquets.

Floral Supply Syndicate
fss.com
Open to the wholesale trade only supplying goods for floral design including floral adhesive, chicken wire, vases, thorn strippers, water tubes, floral waterproof tape, pin frogs, and ribbon.

Floralife
shop.floralife.com
A U.S.-based company supplying post-harvest care and handling products for cut flowers including my favorites for garden roses: D.C.D cleaner, Express Clear 100, QuickDip, Rose Food Clear 300, and Flower Food Express Universal 300 granular packets.

Uline
uline.com
Butcher-Block paper, tissue paper, rubber bands, buckets, garbage cans, sanitation supplies, and assorted tools

IRRIGATION

Dripworks
dripworks.com
Irrigation supplies including drip line, timers and all-in-one kits.

Irrometer
irrometer.com
Commercial soil tensiometers and moisture meters.

UCANR Salinas Valley DIY Soil Tensiometer
ucanr.edu/blogs/blogcore/postdetail.cfm?postnum=30042

TOOLS & SUPPLIES

A.M. Leonard
amleo.com

Farmers Friend
farmersfriend.com

Johnny's Selected Seeds
johnnyseeds.com

SUGGESTED NUTRIENTS FOR ROSE TISSUE

NUTRIENT	LOW	HIGH
N (%)	3.0	5.0
P (%)	0.2	0.3
K (%)	2.0	3.0
Ca (%)	1.0	1.5
Mg (%)	0.25	0.35
Zn (%)	15	50
Mn (ppm)	30	250
Fe (ppm)	50	150
Cu (ppm)	5	15
B (ppm)	30	60

SUGGESTED SOIL PH, EC, AND NUTRIENT LEVELS FOR ROSES

SOIL CHARACTERISTICS	UNITS	LOW	HIGH
PH (ACIDITY/ALKALINITY)	6	6.0	7.5
ECe (ELECTRICAL CONDUCTIVITY)	d5/m	0.5	2.0
NO3-N (NITRATE-N)	ppm	35	150
NH4-N (AMMONIACAL-N)	ppm	0	20
P (PHOSPHORUS)	ppm	5	50
K (POTASSIUM)	ppm	50	300
Ca (CALCIUM)	ppm	40	200
Mg (MAGNESIUM)	ppm	20	100
B (BORON)	ppm	0.1	0.75
Fe (IRON)	ppm	0.3	3.0
Mn (MANGANESE)	ppm	0.2	3.0
Cu (COPPER)	ppm	0.001	0.5
Zn (ZINC)	ppm	0.03	3.0
Mo (MOLYBDENUM)	ppm	0.01	0.10

SOURCE: UNIVERSITY OF CALIFORNIA AGRICULTURE & NATURAL RESOURCES (UCANR)

PRACTICAL INTERPRETATION SOIL MOISTURE CHART FOR SOIL TEXTURES AND CONDITIONS

AVAILABLE SOIL MOISTURE	FEEL OR APPEARANCE OF SOIL			
	COARSE-TEXTURED SOILS	MODERATELY-COARSE-TEXTURED SOILS	MEDIUM-TEXTURED SOILS	FINE AND VERY FINE-TEXTURED SOILS
0 percent	Dry, loose, and single-grained; flows through fingers	Dry and loose; flows through fingers	Powdery dry; in some places slightly crusted, but breakdowns easily into powder	Hard, baked and cracked; has loose crumbs on surface in some places
50 percent or less	Appears to be dry; does not form a ball under pressure	Appears to be dry; does not form a ball under pressure	Somewhat crumbly but holds together under pressure	Somewhat pliable under pressure
50 to 75 percent	Appears to be dry; does not form a ball under pressure	Balls under pressure but seldom holds together	Forms a ball under pressure, somewhat plastic; sticks slightly under pressure	Forms a ball; ribbons out between thumb and forefingers
75 percent to field capacity	Sticks together slightly; may form a very weak ball under pressure	Forms a weak ball that breaks easily; does not stick	Forms ball; very pliable; sticks readily if relatively high in clay	Ribbons out between fingers easily; has a slick feeling
At field capacity (100 percent)	On squeezing, no free water appears on soils but wet outline of ball is left on hand	Same as for coarse-textured soils at field capacity	Same as for coarse-textured soils at field capacity	Same as for coarse-textured soils at field capacity
Above field capacity	Free water appears when soil is bounced in hand	Free water is released with kneading	Free water can be squeezed out	Puddles, free water forms on surface

SOURCE: UNITED STATES DEPARTMENT OF AGRICULTURE (USDA)

MORE ABOUT SOIL TENSION AND HOW TO MEASURE IT

Soil tension is measured as a negative pressure and the device used to measure it is called a tensiometer. Tensiometers indicate the availability of water in the soil. Plant available water, expressed as "kPa" or kilopascal, is the range of soil tension that helps a plant access water from the soil. Generally, it's from about -2kPa to around -40 kPa for most plants. Every plant has their sweet spot for moisture, and garden roses grow best at 10-35 kPa at a depth of 12-24 inches in the root zone.

You can purchase a soil tensioner and a DIY measuring system for your farm or garden or, if you are a flower farmer, you may decide to install a professional monitoring system. Use the soil meter to determine how often to water your roses. In order to get an accurate reading of the water available to your roses, install the soil meter along the water line and drip-line in the rose's root zone. Available water/soil moisture will have the greatest effect on your stem length and overall plant health.

Field-grown garden roses are the healthiest when the soil is maintained at a 10-35 kPa at a depth of 18-24 inches. Check the tensiometer daily and if the tension is above 35 kPa you should apply water. Once that level reaches 10 kPa, shut off the water until it measures just above 35kPa and turn the water on again. If it reaches levels above 35kPa the rose will have drought stress and head towards wilting point.

The higher the number climbs the more stress will occur. I use a tensiometer on my farm daily to guide my irrigation. For flower farmers or Aspiring Rosarians, I know using data from a tensiometer will be a welcome new addition to your rose "tool box".

SOIL RESOURCES

UC Davis California Resource Lab
A healthy rose starts in the ground.
Map your soil in your location with these online interactive Google Maps. Don't forget to get an annual soil test.
casoilresource.lawr.ucdavis.edu/soilweb-apps/

REFERENCES

Dr. Linda Chalker-Scott, PhD
Associate Professor and Extension Specialist, Urban Horticulture
Dr. Chalker-Scott's research in phosphate fertilizers, mulch, and gardening myths are a must-read for any rose gardener or flower farmer.
www.puyallup.wsu.edu/lcs
www.gardenprofessors.com

PEST & DISEASE CONTROL

Do My Own
domyown.com
Large selection of pest- and disease-control products and safety equipment.

GopherX
gopherx.com
Gassing machine to kill gophers, moles, voles, and burrowing pests.

Koppert Biologial Systems
www.koppert.com
Supplier of predatory mites to treat thrips and other beneficial insects for pest control.

National Pesticide
Safety Information Center
www.npsec.us
Information and resource guide on pesticide safety for both organic and non-organic products.

Pesticide Safety - A Reference Manual for Private Applicators University of California Statewide I.P.M Division of Agriculture and Natural Resources Publication 3383
Reference book to learn the proper safety protocols for mixing and applying chemical control measures, both organic and non-organic.

The Rose Doctor: A Key for Diagnosing Problems in the Rose Garden by Gary A. Ritchie PhD
This book is comprehensive guide of the pests and diseases of roses.

WEED CONTROL

Common Weeds of the United States
by the U.S. Department of Agriculture

University of California Agriculture and Natural Resources Weed Photo Gallery
ipm.ucanr.edu/PMG/weeds_intro

University of California Agriculture and Natural Resources How To Solarize Weeds
ipm.ucanr.edu/PMG/PESTNOTES/pn74145

PHOTOGRAPHY BY JILL CARMEL